MECHANICS OF AIRCRAFT STRUCTURES

MECHANICS OF AIRCRAFT STRUCTURES

SECOND EDITION

C. T. Sun

JOHN WILEY & SONS, INC.

Copyright © 2006 by John Wiley & Sons, New York. All rights reserved.

Published by John Wiley & Sons, Inc., Hoboken, New Jersey.
Published simultaneously in Canada.

For general information on our other products and services please contact our Customer Care Department within the United States at (800) 762-2974, outside the United States at (317) 572-3993 or fax (317) 572-4002.

Wiley also publishes its books in a variety of electronic formats. Some content that appears in print may not be available in electronic books. For more information about Wiley products, visit our web site at www.Wiley.com.

Library of Congress Cataloging-in-Publication Data

Sun, C. T. (Chin-Teh), 1939–
    Mechanics of aircraft structures / C. T. Sun.—2nd ed.
        p.   cm.
    Includes index.
        ISBN-13 978-0-471-69966-8 (cloth)
        ISBN-10 0-471-69966-7 (cloth)
        1. Airframes.   2. Fracture mechanics.   I. Title.
    TL671.6.S82   2006
    629.134′31—dc22   2005056952

Printed in the United States of America

10 9 8 7 6 5 4 3

To my wife, Iris, and my children,
Edna, Clifford, and Leslie

CONTENTS

PREFACE

The purpose of the second edition is to correct a number of typographical errors in the first edition, add more examples and problems for the student, and introduce a few new topics, including primary warping, effects of boundary constraints, Saint-Venant's principle, the concept of shear lag, the Timoshenko beam theory, and a brief introduction to the effect of plasticity on fracture. All these additions are direct extensions of the existing contents in the first edition. Consequently, the background-building chapters, Chapters 1 and 2, need no modification. The expansions are concentrated in Chapters 3, 4, and 6 and amount to about a 25 percent increase in the number of pages.

The author is indebted to many students and colleagues for numerous corrections and valuable suggestions. He is indebted also to Dr. G. Huang for his assistance in making many new drawings.

C. T. SUN

PREFACE TO THE FIRST EDITION

This book is intended for junior or senior level aeronautical engineering students with a background in the first course of mechanics of solids. The contents can be covered in a semester at a normal pace.

The selection and presentation of materials in the course of writing this book were greatly influenced by the following developments. First, commercial finite element codes have been used extensively for structural analyses in recent years. As a result, many simplified ad hoc techniques that were important in the past have lost their useful roles in structural analyses. This development leads to the shift of emphasis from the problem-solving drill to better understanding of mechanics, developing the student's ability in formulating the problem, and judging the correctness of numerical results. Second, fracture mechanics has become the most important tool in the study of aircraft structure damage tolerance and durability in the past thirty years. It seems highly desirable for undergraduate students to get some exposure to this important subject, which has traditionally been regarded as a subject for graduate students. Third, advanced composite materials have gained wide acceptance for use in aircraft structures. This new class of materials is substantially different from traditional metallic materials. An introduction to the characteristic properties of these new materials seems imperative even for undergraduate students.

In response to the advent of the finite element method, consistent elasticity approach is employed. Multidimensional stresses, strains, and stress-strain relations are emphasized. Displacement, rather than strain or stress, is used in deriving the governing equations for torsion and bending problems. This approach will help the student understand the relation between simplified structural theories and 3-D elasticity equations.

The concept of fracture mechanics is brought in via the original Griffith's concept of strain energy release rate. Taking advantage of its global nature and its relation to the change of the total strain energies stored in the structure before and after crack extension, the strain energy release rate can be calculated for simple structures without difficulty for junior and senior level students.

The coverage of composite materials consists of a brief discussion of their mechanical properties in Chapter 1, the stress-strain relations for anisotropic solids in Chapter 2, and a chapter (Chapter 8) on analysis of symmetric laminates of composite materials. This should be enough to give the student a background to deal correctly with composites and to avoid regarding a composite as an aluminum alloy with the Young's modulus taken equal to the longitudinal modulus of the composite. Such a brief introduction to composite materials and laminates is by no means sufficient to be used as a substitute for a course (or courses) dedicated to composites.

A classical treatment of elastic buckling is presented in Chapter 7. Besides buckling of slender bars, the postbuckling concept and buckling of structures composed of thin sheets are also briefly covered without invoking an advanced background in solid mechanics. Postbuckling strengths of bars or panels are often utilized in aircraft structures. Exposure, even very brief, to this concept seems justified, especially in view of the mathematics employed, which should be quite manageable for student readers of this book.

The author expresses his appreciation to Mrs. Marilyn Engel for typing the manuscript and to James Chou and R. Sergio Hasebe for making the drawings.

C. T. Sun

MECHANICS OF
AIRCRAFT STRUCTURES

1

CHARACTERISTICS OF AIRCRAFT STRUCTURES AND MATERIALS

1.1 INTRODUCTION

The main difference between aircraft structures and materials and civil engineering structures and materials lies in their weight. The main driving force in aircraft structural design and aerospace material development is to reduce weight. In general, materials with high stiffness, high strength, and light weight are most suitable for aircraft applications.

Aircraft structures must be designed to ensure that every part of the material is used to its full capability. This requirement leads to the use of shell-like structures (monocoque constructions) and stiffened shell structures (semimonocoque constructions). The geometrical details of aircraft structures are much more complicated than those of civil engineering structures. They usually require the assemblage of thousands of parts. Technologies for joining the parts are especially important for aircraft construction.

The size and shape of an aircraft structural component are usually determined based on nonstructural considerations. For instance, the airfoil is chosen according to aerodynamic lift and drag characteristics. Then the solutions for structural problems in terms of global configurations are limited. Often, the solutions resort to the use of special materials developed for applications in aerospace vehicles.

Because of their high stiffness/weight and strength/weight ratios, aluminum and titanium alloys have been the dominant aircraft structural materials for many decades. However, the recent advent of advanced fiber-reinforced composites has changed the outlook. Composites may now achieve weight savings of 30 to 40 percent over aluminum or titanium counterparts. As a result, composites have been used increasingly in aircraft structures. Figure 1.1 shows the key materials on the Boeing-McDonnell-Douglas F/A–18E fighter jet. On the latest Boeing commercial airliner, the 787, composites account for up to 50 percent of structural weight.

1.2   BASIC STRUCTURAL ELEMENTS IN AIRCRAFT STRUCTURE

Major components of aircraft structures are assemblages of a number of basic structural elements, each of which is designed to take a specific type of load, such as axial, bending, or torsional load. Collectively, these elements can efficiently provide the capability for sustaining loads on an airplane. The governing equations for these basic structural elements are introduced in the first course in mechanics of solids. In the following subsections, the governing equations are reviewed briefly and their behavior discussed.

1.2.1   Axial Member

Axial members are used to carry extensional or compressive loads applied in the direction of the axial direction of the member. The resulting stress is uniaxial:

$$\sigma = E\varepsilon \tag{1.1}$$

where E and ε are the Young's modulus and normal strain, respectively, in the loading direction. The total axial force F provided by the member is

$$F = A\sigma = EA\varepsilon \tag{1.2}$$

where A is the cross-sectional area of the member. The quantity EA is termed the axial stiffness of the member, which depends on the modulus of the material and the cross-sectional area of the member. It is obvious that the axial stiffness of axial members cannot be increased (or decreased) by changing the shape of the cross-section. In other words, a circular rod and a channel (see Figs. 1.2a and 1.2b) can carry the same axial load as long as they have the same cross-sectional area.

Axial members are usually slender and are susceptible to buckling failure when subjected to compression. Buckling strength can be increased by increasing the bending stiffness and by shortening the length of the buckle mode. For buckling, the channel section is better since it has higher bending stiffness than the circular section. However, because of the slenderness

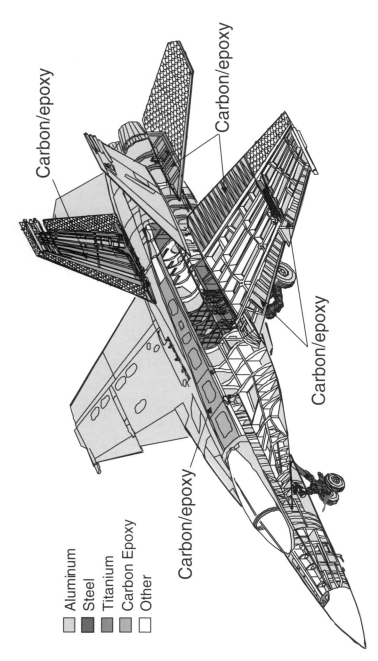

Aluminum
Steel
Titanium
Carbon Epoxy
Other

Carbon/epoxy

Carbon/epoxy

Carbon/epoxy

Carbon/epoxy

Figure 1.1 Materials used on Boeing-McDonnell-Douglas F/A−18E jet. (Courtesy of the Boeing Company.)

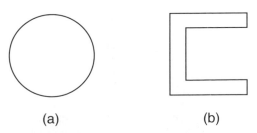

(a)   (b)

Figure 1.2   (a) Circular rod; (b) channel.

of most axial members used in aircraft (such as stringers), the bending stiffness of these members is usually very small and is not sufficient to achieve the necessary buckling strength. In practice, the buckling strength of axial members is enhanced by providing lateral supports along the length of the member with more rigid ribs (in wings) and frames (in fuselage).

1.2.2 Shear Panel

A shear panel is a thin sheet of material used to carry in-plane shear load. Consider a shear panel of uniform thickness t under uniform shear stress τ as shown in Fig. 1.3. The total shear force in the x-direction provided by the panel is given by

$$V_x = \tau t a = G\gamma t a \tag{1.3}$$

where G is the shear modulus, and γ is the shear strain. Thus, for a flat panel, the shear force V_x is proportional to its thickness and the lateral dimension a.

For a curved panel under a state of constant shear stress τ (see Fig 1.4), the resulting shear force of the shear stress on the thin-walled section may be

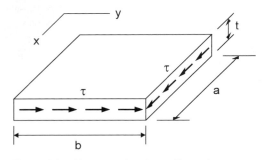

Figure 1.3   Shear panel under uniform shear stress.

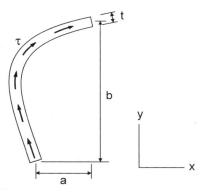

Figure 1.4   Curved panel under a state of constant shear stress.

decomposed into a horizontal component V_x and a vertical component V_y as

$$V_x = \tau t a \tag{1.4}$$

$$V_y = \tau t b \tag{1.5}$$

Thus, the components of the resultant force of the shear stress τ have the relation

$$\frac{V_x}{V_y} = \frac{a}{b}$$

Since this relation does not depend on the contour shape of the section of the panel, a flat panel would be the most efficient (in material usage) in providing a shear force for given values of a and b.

1.2.3   Bending Member (Beam)

A structural member that can carry bending moments is called a **beam**. A beam can also act as an axial member carrying longitudinal tension and compression. According to simple beam theory, bending moment M is related to beam deflection w as

$$M = -EI\frac{d^2 w}{dx^2} \tag{1.6}$$

where EI is the bending stiffness of the beam. The area moment of inertia I depends on the geometry of the cross-section.

Except for pure moment loading, a beam is designed to carry both bending moments and transverse shear forces as the latter usually produce the former. For a beam of a large span/depth ratio, the bending stress is usually more

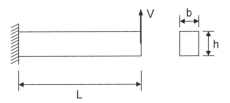

Figure 1.5   Cantilever beam.

critical than the transverse shear stress. This is illustrated by the example of a cantilever beam shown in Fig. 1.5.

It is easy to see that the maximum bending moment and bending stress occur at the fixed root of the cantilever beam. We have

$$\sigma_{max} = \frac{M_{max}(h/2)}{I} = \frac{VL(h/2)}{bh^3/12} = \frac{6VL}{bh^2} \tag{1.7}$$

The transverse shear stress distribution is parabolic over the beam depth with maximum value occurring at the neutral plane, i.e.,

$$\tau_{max} = \frac{3}{2}\frac{V}{bh} \tag{1.8}$$

From the ratio

$$\frac{\sigma_{max}}{\tau_{max}} = \frac{4L}{h} \tag{1.9}$$

it is evident that bending stress plays a more dominant role than transverse shear stress if the span-to-depth ratio is large (as in wing structure). For such beams, attention is focused on optimizing the cross-section to increase bending stiffness.

In the elastic range, bending stress distribution over depth is linear with maximum values at the farthest positions from the neutral axis. The material near the neutral axis is underutilized. Thus, the beam with a rectangular cross-section is not an efficient bending member.

In order to utilize the material to its full capacity, material in a beam must be located as far as possible from the neutral axis. An example is the wide flange beam shown in Fig. 1.6a. Although the bending stress distribution is still linear over the depth, the bending line force (bending stress times the width) distribution is concentrated at the two flanges as shown in Fig. 1.6b because $b \gg t_w$. For simplicity, the small contribution of the vertical web to bending can be neglected.

The transverse shear stress distribution in the wide flange beam is shown in Fig. 1.6c. The vertical web is seen to carry essentially all the transverse shear

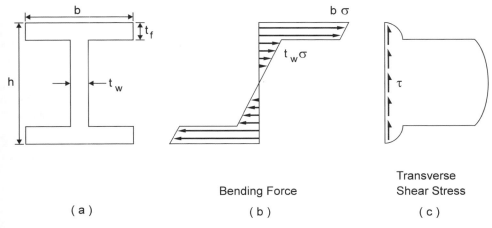

Figure 1.6   (a) Wide-flange beam; (b) bending force distribution; (c) shear stress distribution.

load; its variation over the web is small and can be practically assumed to be constant. For all practical purposes, the wide flange beam can be regarded as two axial members (flanges) connected by a flat shear panel.

1.2.4   Torsion Member

Torque is an important form of load to aircraft structures. In a structural member, torque is formed by shear stresses acting in the plane of the cross-section. Consider a hollow cylinder subjected to a torque T as shown in Fig. 1.7. The torque-induced shear stress τ is linearly distributed along the radial direction. The torque is related to the twist angle θ per unit length as

$$T = GJ\theta \tag{1.10}$$

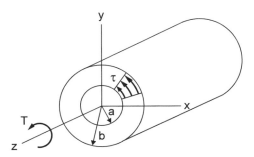

Figure 1.7   Hollow cylinder subjected to a torque.

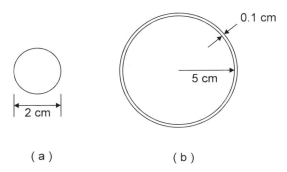

Figure 1.8 Cross-sections of (a) a solid cylinder, and (b) a tube.

where J is the torsional constant. For hollow cylinders, J is equal to the polar moment of inertia of the cross-section, i.e.,

$$J = I_p = \tfrac{1}{2}\pi(b^4 - a^4) = \tfrac{1}{2}\pi(b-a)(b+a)(b^2+a^2) \qquad (1.11)$$

The term GJ is usually referred to as torsional stiffness.

If the wall thickness $t = b - a$ is small compared with the inner radius, then an approximate expression of J is given by

$$J = 2t\pi\bar{r}^3 \qquad (1.12)$$

where $\bar{r} = (a + b)/2$ is the average value of the outer and inner radii. Thus, for a thin-walled cylinder, the torsional stiffness is proportional to the 3/2 power of the area ($\pi\bar{r}^2$) enclosed by the wall.

Note that the material near the inner cavity in a thick-walled cylinder is underutilized. It is obvious that a thin-walled tube would be more efficient for torques than a solid cylinder or a thick-walled hollow cylinder. Figure 1.8 shows the cross-sections of a solid cylinder (Fig. 1.8a) and a tube (Fig. 1.8b), both having the same amount of material. Using (1.11) or (1.12), it is easy to show that the torsional stiffness of the tube is almost 50 times that of the solid cylinder. This example illustrates that a thin-walled structure can be made into a very efficient torsion member.

1.3 WING AND FUSELAGE

The wing and fuselage are the two major airframe components of an airplane. The horizontal and vertical tails bear close resemblance to the wing. Hence, these two components are taken for discussion to exemplify the principles of structural mechanics employed in aircraft structures.

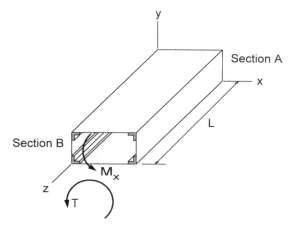

Figure 1.9   Box beam.

1.3.1   Load Transfer

Wing and fuselage structures consist of a collection of basic structural elements. Each component, as a whole, acts like a beam and a torsion member. For illustrative purposes, let us consider the box beam shown in Fig. 1.9. The box beam consists of stringers (axial members) that are located at the maximum allowable distance from the neutral axis to achieve the most bending capability, and the thin skin (shear panel), which encloses a large area to provide a large torque capability. The design of Fig. 1.9 would be fine if the load is directly applied in the form of global torque T and bending moment M_x. In reality, aircraft loads are in the form of air pressure (or suction) on the skin, concentrated loads from the landing gear, power plants, passenger seats, etc. These loads are to be "collected" locally and transferred to the major load-carrying members. Without proper care, these loads may produce excessive local deflections that are not permissible from aerodynamic considerations.

Using the box beam of Fig. 1.9 as an example, we assume that a distributed air pressure is applied on the top and bottom surfaces of the beam. The skin (shear panel) is thin and has little bending stiffness to resist the air pressure. To avoid incurring large deflections in the skin, longitudinal stringers (stiffeners) can be added, as shown in Fig. 1.10, to pick up the air loads. These stiffeners are usually slender axial members with a moderate amount of bending stiffness. Therefore, the transverse loads picked up by the stiffeners must be transferred "quickly" to more rigid ribs or frames at sections A and B (see Fig. 1.9) to avoid excessive deflections. The ribs collect all transverse loads from the stiffeners and transfer them to the two wide-flange beams (spars) that are designed to take transverse shear loads.

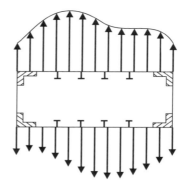

Figure 1.10   Longitudinal stringers in a box beam.

The local-to-global load transfer is thus complete. Note that besides serving as a local load distributor, the stiffeners also contribute to the total bending capability of the box beam.

1.3.2   Wing Structure

The main function of the wing is to pick up the air and power plant loads and transmit them to the fuselage. The wing cross-section takes the shape of an airfoil, which is designed based on aerodynamic considerations. The wing as a whole performs the combined function of a beam and a torsion member. It consists of axial members in stringers, bending members in spars and shear panels in the cover skin and webs of spars. The spar is a heavy beam running spanwise to take transverse shear loads and spanwise bending. It is usually composed of a thin shear panel (the web) with a heavy cap or flange at the top and bottom to take bending. A typical spar construction is depicted in Fig. 1.11. A multiple-spar wing construction is shown in Fig. 1.1.

Wing ribs are planar structures capable of carrying in-plane loads. They are placed chordwise along the wing span. Besides serving as load redistributers, ribs also hold the skin stringer to the designed contour shape. Ribs

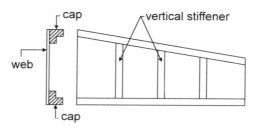

Figure 1.11   Typical spar construction.

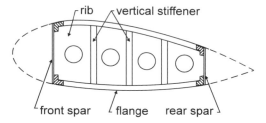

Figure 1.12   Typical rib construction.

reduce the effective buckling length of the stringers (or the stringer-skin system) and thus increase their compressive load capability. Figure 1.12 shows a typical rib construction. Note that the rib is supported by spanwise spars.

The cover skin of the wing together with the spar webs form an efficient torsion member. For subsonic airplanes, the skin is relatively thin and may be designed to undergo postbuckling. Thus, the thin skin can be assumed to make no contribution to bending of the wing box, and the bending moment is taken by spars and stringers. Figure 1.13 presents two typical wing cross-sections for two-spar subsonic aircraft. One type (Fig. 1.13a) consists only of spars (the concentrated flange type) to take bending. The other type (the distributed flange type, Fig. 1.13b) uses both spars and stringers to take bending.

Supersonic airfoils are relatively thin compared with subsonic airfoils. To withstand high surface air loads and to provide additional bending capability of the wing box structure, thicker skins are often necessary. In addition, to increase structural efficiency, stiffeners can be manufactured (either by forging or machining) as integral parts of the skin.

1.3.3 Fuselage

Unlike the wing, which is subjected to large distributed air loads, the fuselage is subjected to relatively small air loads. The primary loads on the fuselage

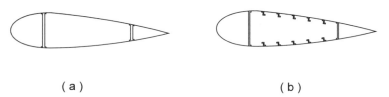

(a)   (b)

Figure 1.13   Typical two-spar wing cross-sections for subsonic aircraft: (a) spars only; (b) spars and stringers.

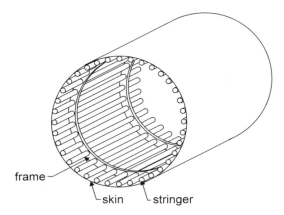

Figure 1.14 Fuselage structure.

include large concentrated forces from wing reactions, landing gear reactions, and pay loads. For airplanes carrying passengers, the fuselage must also withstand internal pressures. Because of internal pressure, the fuselage often has an efficient circular cross-section. The fuselage structure is a semimonocoque construction consisting of a thin shell stiffened by longitudinal axial elements (stringers and longerons) supported by many transverse frames or rings along its length; see Fig. 1.14. The fuselage skin carries the shear stresses produced by torques and transverse forces. It also bears the hoop stresses produced by internal pressures. The stringers carry bending moments and axial forces. They also stabilize the thin fuselage skin.

Fuselage frames often take the form of a ring. They are used to maintain the shape of the fuselage and to shorten the span of the stringers between supports in order to increase the buckling strength of the stringer. The loads on the frames are usually small and self-equilibrated. Consequently, their constructions are light. To distribute large concentrated forces such as those from the wing structure, heavy bulkheads are needed.

Figure 1.15 shows the fuselage of a Boeing 777 under construction.

1.4 AIRCRAFT MATERIALS

Traditional metallic materials used in aircraft structures are aluminum, titanium, and steel alloys. In the past three decades, applications of advanced fiber composites have rapidly gained momentum. To date, some new commercial jets, such as the Boeing 787, already contain composite materials up to 50 percent of their structural weight.

Figure 1.15 Fuselage of a Boeing 777 under construction. (Courtesy of the Boeing Company.)

Selection of aircraft materials depends on many considerations which can, in general, be categorized as cost and structural performance. Cost includes initial material cost, manufacturing cost and maintenance cost. The key material properties that are pertinent to maintenance cost and structural performance are

- Density (weight)
- Stiffness (Young's modulus)
- Strength (ultimate and yield strengths)
- Durability (fatigue)
- Damage tolerance (fracture toughness and crack growth)
- Corrosion

Seldom is a single material able to deliver all desired properties in all components of the aircraft structure. A combination of various materials is often necessary. Table 1.1 lists the basic mechanical properties of some metallic aircraft structural materials.

Steel Alloys   Among the three metallic materials, steel alloys have highest densities, and are used only where high strength and high yield stress are critical. Examples include landing gear units and highly loaded fittings. The high strength steel alloy 300 M is commonly used for landing gear components. This steel alloy has a strength of 1.9 GPa (270 ksi) and a yield stress of 1.5 GPa (220 ksi).

TABLE 1.1   Mechanical properties of metals at room temperature in aircraft structures

	Property[a]				
Material	E GPa (msi)	ν	σ_u MPa (ksi)	σ_Y MPa (ksi)	ρ g/cm^3 (lb/in^3)
Aluminum					
2024-T3	72 (10.5)	0.33	449 (65)	324 (47)	2.78 (0.10)
7075-T6	71 (10.3)	0.33	538 (78)	490 (71)	2.78 (0.10)
Titanium					
Ti-6Al-4V	110 (16.0)	0.31	925 (134)	869 (126)	4.46 (0.16)
Steel					
AISI4340	200 (29.0)	0.32	1790 (260)	1483 (212)	7.8 (0.28)
300 M	200 (29.0)	0.32	1860 (270)	1520 (220)	7.8 (0.28)

[a] σ_u = tensile ultimate stress; σ_Y = tensile yield stress.

Besides being heavy, steel alloys are generally poor in corrosion resistance. Components made of these alloys must be plated for corrosion protection.

Aluminum Alloys   Aluminum alloys have played a dominant role in aircraft structures for many decades. They offer good mechanical properties with low weight. Among the aluminum alloys, the 2024 and 7075 alloys are perhaps the most used. The 2024 alloys (2024-T3, T42) have excellent fracture toughness and slow crack growth rate as well as good fatigue life. The code number following T for each aluminum alloy indicates the heat treatment process. The 7075 alloys (7075-T6, T651) have higher strength than the 2024 but lower fracture toughness. The 2024-T3 is used in the fuselage and lower wing skins, which are prone to fatigue due to applications of cyclic tensile stresses. For the upper wing skins, which are subjected to compressive stresses, fatigue is less of a problem, and 7075-T6 is used.

The recently developed aluminum lithium alloys offer improved properties over conventional aluminum alloys. They are about 10 percent stiffer and 10 percent lighter and have superior fatigue performance.

Titanium Alloys   Titanium such as Ti-6Al-4V (the number indicates the weight percentage of the alloying element) with a density of 4.5 g/cm^3 is lighter than steel (7.8 g/cm^3) but heavier than aluminum (2.7 g/cm^3). See Table 1.1. Its ultimate and yield stresses are almost double those of aluminum 7075-T6. Its corrosion resistance in general is superior to both steel and aluminum alloys. While aluminum is usually not for applications above 350°F, titanium, on the other hand, can be used continuously up to 1000°F.

Titanium is difficult to machine, and thus the cost of machining titanium parts is high. Near net shape forming is an economic way to manufacture titanium parts. Despite its high cost, titanium has found increasing use in military aircraft. For instance, the F-15 contained 26 percent (structural weight) titanium.

Fiber-Reinforced Composites   Materials made into fiber forms can achieve significantly better mechanical properties than their bulk counterparts. A notable example is glass fiber versus bulk glass. The tensile strength of glass fiber can be two orders of magnitude higher than that of bulk glass. In this century, fiber science has made gigantic strides, and many high-performance fibers have been introduced. Listed in Table 1.2 are the mechanical properties of some high-performance manufactured fibers.

Fibers alone are not suitable for structural applications. To utilize the superior properties of fibers, they are embedded in a matrix material that holds the fibers together to form a solid body capable of carrying complex loads.

TABLE 1.2   Mechanical properties of fibers

	Property		
Material	E GPa (msi)	σ_u GPa (ksi)	ρ g/cm^3
E-glass	77.0 (11)	2.50 (350)	2.54
S-glass	85.0 (12)	3.50 (500)	2.48
Silicon carbide (Nicalon)	190.0 (27)	2.80 (400)	2.55
Carbon (Hercules AS4)	240.0 (35)	3.60 (510)	1.80
Carbon (Hercules HMS)	360.0 (51)	2.20 (310)	1.80
Carbon (Toray T300)	240.0 (35)	3.50 (500)	1.80
Boron	385.0 (55)	3.50 (500)	2.65
Kevlar-49 (Aramid)	130.0 (18)	2.80 (400)	1.45
Kevlar-29	65.0 (9.5)	2.80 (400)	1.45

Matrix materials that are currently used for forming composites include three major categories: polymers, metals, and ceramics. The resulting composites are usually referred to as polymer matrix composites (PMCs), metal matrix composites (MMCs), and ceramic matrix composites (CMCs). Table 1.3 presents properties of a list of composites. The range of service temperature of a composite is often determined by its matrix material. Polymer matrix composites are usually for lower temperature (less than 300°F) applications, and ceramic matrix composites are intended for applications in hot (higher than 1500°F) environments, such as jet engines.

Fiber composites are stiff, strong, and light and are thus most suitable for aircraft structures. They are often used in the form of laminates that consist of a number of unidirectional laminae with different fiber orientations to provide multidirectional load capability. Composite laminates have excellent fatigue life, damage tolerance, and corrosion resistance. Laminate constructions offer the possibility of tailoring fiber orientations to achieve optimal structural performance of the composite structure.

TABLE 1.3   Longitudinal mechanical properties of fiber composites

		Property		
Material	Type	E GPa (msi)	σ_u GPa (ksi)	ρ g/cm^3
Carbon/epoxy	T300/5208	140.0 (20)	1.50 (210)	1.55
	IM6/3501-6	177.0 (25.7)	2.86 (414)	1.55
	AS4/3501-6	140.0 (20)	2.10 (300)	1.55
Boron/aluminum	B/Al 2024	210.0 (30)	1.50 (210)	2.65
Glass/epoxy	S2 Glass/epoxy	43.0 (6.2)	1.70 (245)	1.80
Aramid/epoxy	Kev 49/epoxy	70.0 (10)	1.40 (200)	1.40

PROBLEMS

1.1 The beam of a rectangular thin-walled section (i.e., t is very small) is designed to carry both bending moment M and torque T. If the total wall contour length $L = 2(a + b)$ (see Fig. 1.16) is fixed, find the optimum b/a ratio to achieve the most efficient section if $M = T$ and $\sigma_{\text{allowable}} = 2\tau_{\text{allowable}}$. Note that for closed thin-walled sections such as the one in Fig. 1.16, the shear stress due to torsion is

$$\tau = \frac{T}{2abt}$$

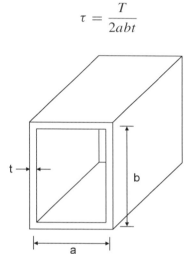

Figure 1.16   Closed thin-walled section.

1.2 Do Problem 1.1 with $M = \alpha T$ where $\alpha = 0$ to ∞.

1.3 The dimensions of a steel (300 M) I-beam are $b = 50$ mm, $t = 5$ mm, and $h = 200$ mm (Fig. 1.17). Assume that t and h are to be fixed for an aluminum (7075-T6) I-beam. Find the width b for the aluminum beam so that its bending stiffness EI is equal to that of the steel beam. Compare the weights-per-unit length of these two beams. Which is more efficient weightwise?

1.4 Use AS4/3501-6 carbon/epoxy composite to make the I-beam as stated in Problem 1.3. Compare its weight with that of the aluminum beam.

1.5 Derive the relations given by (1.4) and (1.5).

1.6 The sign convention (positive directions of resultants) used in the beam theory depends on the coordinate system chosen. Consider the moment–curvature relation

$$M = -EI\frac{d^2 w}{dx^2}$$

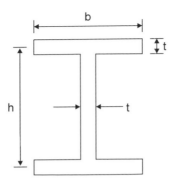

Figure 1.17 Dimensions of the cross-section of an I-beam.

in reference to the coordinate system shown in Fig. 1.18. If w is regarded as a positive displacement (or deflection) in the positive y-direction, find the positive direction of the bending moment. State the reason.

Figure 1.18 Coordinate system for a beam.

1.7 Compare the load-carrying capabilities of two beams having the respective cross-sections shown in Fig. 1.19. Use bending rigidity as the criterion for comparison. It is given that $a = 4$ cm, $t = 0.2$ cm, and the two cross-sections have the same area.

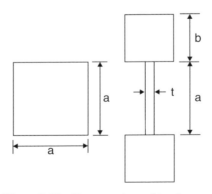

Figure 1.19 Cross-sections of two beams.

2

INTRODUCTION
TO ELASTICITY

2.1 CONCEPT OF DISPLACEMENT

Consider a material point P at the position $\mathbf{x}(x, y, z)$ before deformation (see Fig. 2.1). After deformation, P moves to a new position $P'(x', y', z')$. The change of position during deformation, which is measured in terms of the displacement vector $\mathbf{u}$, has three components: u, v, and w in the x, y, and z directions, respectively. The new location of the point (x, y, z) after deformation is given by

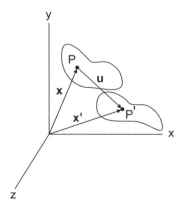

Figure 2.1 Displacement of material point P after deformation.

$$x' = x + u \quad \text{or} \quad u = x' - x$$
$$y' = y + v \quad\quad v = y' - y \tag{2.1}$$
$$z' = z + w \quad\quad w = z' - z$$

Thus, the deformed configuration is uniquely defined if the displacement components u, v, and w are given everywhere in the body of interest.

Consider an axial member [i.e., a one-dimensional (1-D) body] of original length L_0. Assume the axial strain to be uniform in the member. Then the axial strain everywhere in the member is calculated by

$$\varepsilon = \frac{\Delta L}{L_0} \tag{2.2}$$

where ΔL is the total elongation of the member. The elongation ΔL can be regarded as the difference in displacement $u_1 = u(x_1)$ at the right end and $u_0 = u(x_0)$ at the left end (see Fig. 2.2), i.e.,

$$\Delta L = u_1 - u_0$$

The function $u(x) = u_0 + \varepsilon_0(x - x_0)$ gives the axial displacement at any point x in the axial member.

If the strain is not uniform, then (2.2) gives an average strain. To determine the strain at a point, a small segment $L_0 = \Delta x$ must be considered. Consider two points x_0 and $x_0 + \Delta x$ that are separated by a small distance Δx. Let the displacements at these two points be

$$u_0 = u(x_0)$$

and

$$u_1 = u(x_0 + \Delta x)$$

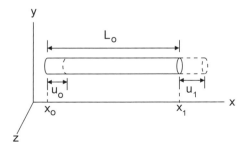

Figure 2.2   Elongation.

respectively. The difference in displacement between these two points is

$$\Delta u = u_1 - u_0 = u(x_0 + \Delta x) - u(x_0) \tag{2.3}$$

which can also be regarded as the elongation of the material between these two points. The axial strain in this segment (or at point x_0) is defined as

$$\varepsilon = \lim_{\Delta x \to 0} \frac{\Delta u}{\Delta x} = \frac{du}{dx} \tag{2.4}$$

Thus, axial strain can be obtained from the derivative of the displacement function.

If a rod is subjected to a uniform tension and $\varepsilon = \varepsilon_0 = $ constant, then

$$\frac{du}{dx} = \varepsilon_0, \qquad x_0 \leq x \leq x_0 + L_0$$

Integrate the equation above to obtain

$$u = \varepsilon_0 x + C$$

Let $u(x_0) = u_0$; then, from the equation above, $C = u_0 - \varepsilon_0 x_0$, and the displacement function is given by

$$u = \varepsilon_0 (x - x_0) + u_0 \tag{2.5}$$

2.2 STRAIN

Consider two points P and Q in a solid body. The coordinates of P and Q are (x, y, z) and $(x + \Delta x, y, z)$, respectively. The distance between the two points before deformation is Δx (see Fig. 2.3).

After deformation, let the displacement of P in the x-direction be $u = u(x, y, z)$ and of Q be $u' = u(x + \Delta x, y, z)$. The new distance between these

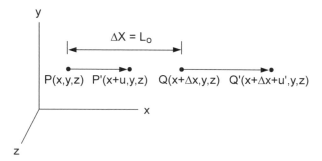

Figure 2.3   Neighboring points P and Q in a solid body.

two points (P' and Q') in the x-direction after deformation is

$$(x + \Delta x + u') - (x + u) = \Delta x + \Delta u \qquad (2.6)$$

where $\Delta u \equiv u' - u$ is the *change of length* in the x-direction for material connecting P and Q after deformation. The strain is defined just as in an axial member:

$$\varepsilon_{xx} = \lim_{\Delta x \to 0} \frac{\Delta u}{\Delta x} = \frac{\partial u}{\partial x} \qquad (2.7)$$

This is the x-component of the normal strain, which measures the *deformation in the x-direction* at a point (x, y, z).

Similarly, the y and z components of the normal strain at the point are given by

$$\varepsilon_{yy} = \frac{\partial v}{\partial y} \qquad (2.8)$$

and

$$\varepsilon_{zz} = \frac{\partial w}{\partial z} \qquad (2.9)$$

respectively. Comparing the strain component ε_{xx} with the strain in the 1-D case (or in an axial member), we may interpret ε_{xx} as the elongation per unit length of an "infinitesimal" axial element of the material at a point (x, y, z) in the x-direction. Similar interpretations can be given to ε_{yy} and ε_{zz}.

The three normal strain components are not sufficient to describe a general state of deformation in a 3-D body. Additional shear strain components are needed to describe the distortional deformation.

For simplicity, let us consider a 2-D case. Let P, Q, R be three neighboring points all lying on the $x-y$ plane as shown in Fig. 2.4. Let P', Q', and R'

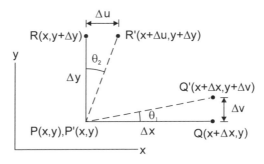

Figure 2.4   Rotations of material line elements in the $x-y$ plane.

be the corresponding positions after deformation. For no loss of generality, we assume that P does not move. In addition, we assume that the infinitesimal elements $\overline{PQ}$ and $\overline{PR}$ do not experience any elongation. Thus, the positions of P' and Q' (see Fig. 2.4) are given by

$$Q': \quad (x + \Delta x, y + \Delta v)$$
$$R': \quad (x + \Delta u, y + \Delta y)$$

For Q', the displacement increment Δv is

$$\Delta v = \underbrace{v(x + \Delta x, y)}_{\text{displacement at } Q} - \underbrace{v(x, y)}_{\text{displacement at } P} \tag{2.10}$$

Similarly for R', the displacement increment Δu can be written as

$$\Delta u = u(x, y + \Delta y) - u(x, y) \tag{2.11}$$

The rotations θ_1 and θ_2 of elements $\overline{PQ}$ and $\overline{PR}$ are assumed to be small and are given by

$$\theta_1 = \lim_{\Delta x \to 0} \frac{\Delta v}{\Delta x} = \frac{\partial v}{\partial x}$$

and

$$\theta_2 = \lim_{\Delta x \to 0} \frac{\Delta u}{\Delta y} = \frac{\partial u}{\partial y}$$

respectively. The total change of angle between $\overline{PQ}$ and $\overline{PR}$ after deformation is defined as the shear strain component in the x–y plane:

$$\gamma_{xy} = \gamma_{yx} \equiv \theta_1 + \theta_2 = \frac{\partial v}{\partial x} + \frac{\partial u}{\partial y} \tag{2.12}$$

Similar shear strain components in the y–z plane and x–z plane are defined as

$$\gamma_{zy} = \gamma_{yz} \equiv \frac{\partial w}{\partial y} + \frac{\partial v}{\partial z} \tag{2.13}$$

$$\gamma_{zx} = \gamma_{xz} \equiv \frac{\partial w}{\partial x} + \frac{\partial u}{\partial z} \tag{2.14}$$

Thus, a general state of deformation at a point in a solid is described by three normal strain components ε_{xx}, ε_{yy}, ε_{zz} and three shear strain components γ_{xy}, γ_{yz}, γ_{xz}.

Rigid Body Motion   If a body undergoes a displacement without inducing strains in the body, then the motion is a rigid body motion. For instance, the displacements

$$u = u_0 = \text{constant}$$

$$v = v_0 = \text{constant}$$

$$w = w_0 = \text{constant}$$

represent a rigid body translational motion and do not yield any strains.

Another rigid body motion is the rigid body rotation. The following displacements represent a rigid body rotation in the $x-y$ plane.

$$u = -\alpha y$$

$$v = \alpha x \qquad\qquad (2.15)$$

$$w = 0$$

It is easy to verify that no strains are associated with the displacement field above.

Example 2.1   Simple Shear   Consider a 2-D body (a unit square $ABCD$) in the $x-y$ plane as shown in Fig. 2.5. After deformation, the four corner points move to A', B', C', and D', respectively. Assume that the displacement field is given by

$$u = 0.01 y$$

$$v = 0.015 x \qquad\qquad (a)$$

Using (2.1) the new position of point A after deformation is given by

$$x' = 0 + u|_{x=0, y=1} = 0.01$$

$$y' = 1 + v|_{x=0, y=1} = 1 + 0 = 1$$

New coordinates of A': $(0.01, 1)$

Similarly, we obtain the new positions of B, C, and D.

$$B': \quad (1.01, 1.015)$$

$$C': \quad (0, 0)$$

$$D': \quad (1, 0.015)$$

Since the deformed configuration is linear in x and y, it can be determined from the new positions A', B', C', and D' as shown by the dashed lines in Fig. 2.5.

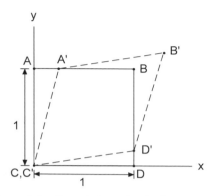

Figure 2.5 Shear deformation in the $s-y$ plane.

The strains corresponding to the displacements given by (a) are

$$\varepsilon_{xx} = \frac{\partial u}{\partial x} = 0, \qquad \varepsilon_{yy} = \frac{\partial v}{\partial y} = 0$$

$$\gamma_{xy} = \frac{\partial v}{\partial x} + \frac{\partial u}{\partial y} = 0.015 + 0.01 = 0.025$$

This is a simple shear deformation.

The following displacement field,

$$u = 0.01y$$

$$v = -0.01x$$

which does not yield any nonvanishing strain component, is a rigid body rotation.

2.3 STRESS

For an axial member, the force is always parallel to the member, and the stress is defined as

$$\sigma = \frac{P}{A} \tag{2.16}$$

where A is the cross-sectional area. If $A = 1$ unit area, then $\sigma = P$.

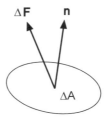

Figure 2.6   Total resultant force on an area.

The concept of stress can easily be extended to 3-D bodies subjected to loads applied in arbitrary directions. Consider an infinitesimal plane surface of area ΔA with a unit normal vector **n**. The total resultant force acting on this area is $\Delta \mathbf{F}$ (force is a vector; see Fig. 2.6). The **stress vector t** is defined as

$$\mathbf{t} = \lim_{\Delta A \to 0} \frac{\Delta \mathbf{F}}{\Delta A} \tag{2.17}$$

Thus, **t** can be considered as the force per unit area acting on the given plane surface.

Consider the special plane surface with the unit normal vector parallel to the x-axis. On this face, the stress vector **t** has three components, which are denoted by σ_{xx}, τ_{xy}, τ_{xz} as shown in Fig. 2.7. Similarly, on the y and z faces the force intensities are given by the components of the respective

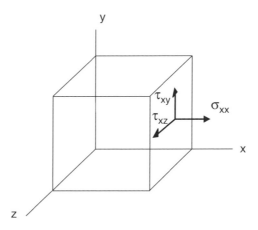

Figure 2.7   Stress vector with three components on the x-face.

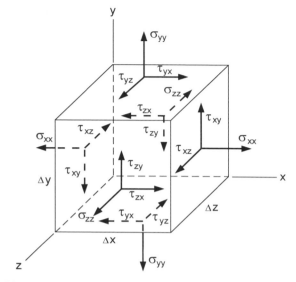

Figure 2.8   Infinitesimal solid element under uniform stress.

stress vectors as

$$\sigma_{yy}, \ \tau_{yx}, \ \tau_{yz} \quad \text{and} \quad \sigma_{zz}, \ \tau_{zx}, \ \tau_{zy}.$$

Consider an infinitesimal solid element under a state of uniform stress. The stress components on the six faces of this element are shown in Fig. 2.8. Since the body is in equilibrium, the six equations of equilibrium must be satisfied, i.e.,

$$\sum F_x = 0, \quad \sum F_y = 0, \quad \sum F_z = 0 \qquad (2.18)$$

$$\sum M_x = 0, \quad \sum M_y = 0, \quad \sum M_z = 0 \qquad (2.19)$$

The force equations (2.18) are obviously satisfied automatically. To satisfy the moment equation (2.19), the following relations among the shear stress components are necessary.

$$\tau_{xy} = \tau_{yx}, \qquad \tau_{yz} = \tau_{zy}, \qquad \tau_{xz} = \tau_{zx} \qquad (2.20)$$

Thus, only six stress components are independent, including three normal stress components σ_{xx}, σ_{yy}, σ_{zz} and three shear stress components, say, τ_{yz}, τ_{xz}, τ_{xy}.

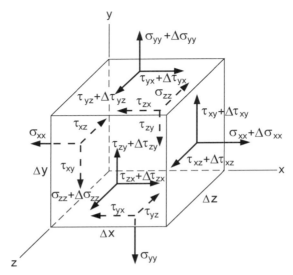

Figure 2.9   Stress components acting on the faces of the element under a nonuniform state of stress.

2.4   EQUATIONS OF EQUILIBRIUM IN A NONUNIFORM STRESS FIELD

Consider an infinitesimal element $\Delta x \times \Delta y \times \Delta z$ in which the stress field is not uniform. The stress components acting on the faces of the element are shown in Fig. 2.9.

If the element is in equilibrium, then the six equations of equilibrium, (2.18) and (2.19), must be satisfied. Consider one of the equations of equilibrium, say, $\sum F_x = 0$. We have

$$(\sigma_{xx} + \Delta\sigma_{xx})\, \Delta y\, \Delta z - (\sigma_{xx})\, \Delta y\, \Delta z$$
$$+ (\tau_{yx} + \Delta\tau_{yx})\, \Delta x\, \Delta z - (\tau_{yx})\, \Delta x\, \Delta z$$
$$+ (\tau_{zx} + \Delta\tau_{zx})\, \Delta x\, \Delta y - (\tau_{zx})\, \Delta x\, \Delta y$$
$$= 0$$

Dividing the equation above by $\Delta x\, \Delta y\, \Delta z$, we obtain

$$\frac{\Delta\sigma_{xx}}{\Delta x} + \frac{\Delta\tau_{yx}}{\Delta y} + \frac{\Delta\tau_{zx}}{\Delta z} = 0$$

Taking the limit $\Delta x,\ \Delta y,\ \Delta z \to 0$, the equilibrium equation above becomes

$$\frac{\partial\sigma_{xx}}{\partial x} + \frac{\partial\tau_{yx}}{\partial y} + \frac{\partial\tau_{zx}}{\partial z} = 0 \tag{2.21}$$

Similarly, equations $\sum F_y = 0$ and $\sum F_z = 0$ lead to

$$\frac{\partial \tau_{xy}}{\partial x} + \frac{\partial \sigma_{yy}}{\partial y} + \frac{\partial \tau_{zy}}{\partial z} = 0 \tag{2.22}$$

and

$$\frac{\partial \tau_{xz}}{\partial x} + \frac{\partial \tau_{yz}}{\partial y} + \frac{\partial \sigma_{zz}}{\partial z} = 0 \tag{2.23}$$

respectively.

It can easily be verified that the moment equations $\sum M_x = \sum M_y = \sum M_z = 0$ lead to

$$\tau_{xy} = \tau_{yx}, \quad \tau_{yz} = \tau_{zy}, \quad \tau_{xz} = \tau_{zx} \tag{2.24}$$

which are identical to the relation given by (2.20).

Equations (2.21)–(2.24) are the equilibrium equations of a point in a body. If a body is in equilibrium, the stress field must satisfy these equations everywhere in the body.

Consider a two-dimensional state of stress so that $\sigma_{xx} \neq 0$, $\sigma_{yy} \neq 0$, $\tau_{xy} \neq 0$ and other stress components vanish. Consider the wedge shown in Fig. 2.10. The unit normal vector to the inclined surface is **n**, and the stress vector (force per unit area) acting on this surface is **t**. From the equilibrium equations $\sum F_x = 0$ and $\sum F_y = 0$ for the wedge, we obtain

$$\begin{aligned} t_x \, \Delta s &= \sigma_{xx} \, \Delta y + \tau_{yx} \, \Delta x \\ t_y \, \Delta s &= \tau_{xy} \, \Delta y + \sigma_{yy} \, \Delta x \end{aligned} \tag{2.25}$$

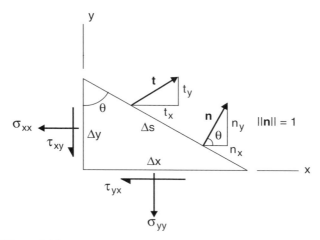

Figure 2.10 Two-dimensional state of stress in a wedge element.

By noting

$$\frac{\Delta y}{\Delta s} = \cos \theta = n_x$$
$$\frac{\Delta x}{\Delta s} = \sin \theta = n_y \tag{2.26}$$

(2.25) can be expressed in the form

$$t_x = \sigma_{xx} n_x + \tau_{yx} n_y$$
$$t_y = \tau_{xy} n_x + \sigma_{yy} n_y \tag{2.27}$$

Equations (2.27) can be expressed in matrix form as

$$\begin{bmatrix} \sigma_{xx} & \tau_{xy} \\ \tau_{xy} & \sigma_{yy} \end{bmatrix} \begin{Bmatrix} n_x \\ n_y \end{Bmatrix} = \begin{Bmatrix} t_x \\ t_y \end{Bmatrix} \tag{2.28}$$

where the relation $\tau_{yx} = \tau_{xy}$ has been invoked.

Using the same method, one can easily derive the equations for the three-dimensional case with the result

$$\begin{bmatrix} \sigma_{xx} & \tau_{xy} & \tau_{xz} \\ \tau_{xy} & \sigma_{yy} & \tau_{yz} \\ \tau_{xz} & \tau_{yz} & \sigma_{zz} \end{bmatrix} \begin{Bmatrix} n_x \\ n_y \\ n_z \end{Bmatrix} = \begin{Bmatrix} t_x \\ t_y \\ t_z \end{Bmatrix} \tag{2.29}$$

Symbolically, (2.29) can be written as

$$[\sigma]\{n\} = \{t\} \tag{2.30}$$

Equation (2.30) indicates that the stress matrix $[\sigma]$ can be viewed as a transformation matrix that transforms the unit normal vector $\{n\}$ into the stress vector $\{t\}$, which acts on the surface with the unit normal $\{n\}$.

Example 2.2   Assume that the uniform state of stress in the wedge-shaped body shown in Fig. 2.11 is

$$[\sigma] = \begin{bmatrix} 2 & 2 & 0 \\ 2 & 2 & 0 \\ 0 & 0 & 1 \end{bmatrix} \qquad \text{MPa}$$

The unit vector **n** normal to the inclined face is given by

$$n_x = n_y = n_z = \frac{1}{\sqrt{3}}$$

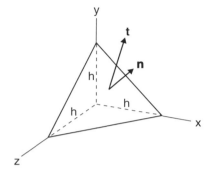

Figure 2.11   Wedge-shaped body.

By using (2.29), the components of the stress vector **t** (force per unit area) are obtained as

$$t_x = \sigma_{xx}n_x + \tau_{xy}n_y + \tau_{xz}n_z = \frac{5}{\sqrt{3}}$$

$$t_y = \tau_{xy}n_x + \sigma_{yy}n_y + \tau_{yz}n_z = \frac{5}{\sqrt{3}} \qquad \text{MPa}$$

$$t_z = \tau_{xz}n_x + \tau_{yz}n_y + \sigma_{zz}n_z = \frac{1}{\sqrt{3}}$$

The total force **F** acting on the inclined face is $A\mathbf{t}$, where A is the area of the inclined face.

2.5   PRINCIPAL STRESS

If we are interested in finding surfaces for which **t** is parallel to **n**, i.e.,

$$\{t\} = \sigma\{n\} \tag{2.31}$$

where σ is a scalar, then (2.30) yields a typical eigenvalue problem:

$$[\sigma]\{n\} = \{t\} = \sigma\{n\} \tag{2.32}$$

or

$$([\sigma] - \sigma[I])\{n\} = 0 \tag{2.33}$$

where $[I]$ is the identity matrix. In order for (2.34) to have a nontrivial solution for $\{n\}$, we require that

$$\|[\sigma] - \sigma[I]\| = 0$$

or, explicitly,

$$\begin{vmatrix} \sigma_{xx} - \sigma & \tau_{xy} & \tau_{xz} \\ \tau_{xy} & \sigma_{yy} - \sigma & \tau_{yz} \\ \tau_{xz} & \tau_{yz} & \sigma_{zz} - \sigma \end{vmatrix} = 0 \qquad (2.34)$$

Expanding the determinantal equation (2.34) yields a cubic equation in σ. Since $[\sigma]$ is real and symmetric, there are three real roots, say σ_1, σ_2, σ_3 (see any book on linear algebra or matrix theory for the proof). The corresponding eigenvectors, $\{n^{(1)}\}$, $\{n^{(2)}\}$, $\{n^{(3)}\}$, can be shown to be mutually orthogonal. These three directions are called **principal directions** of stress, and σ_1, σ_2, and σ_3 are the corresponding **principal stresses**.

On the surfaces perpendicular to these directions, we have according to (2.31),

$$\{t^{(i)}\} = \sigma_i \{n^{(i)}\}, \qquad i = 1, 2, 3 \qquad (2.35)$$

This says that on the surface with the unit normal $\{n^{(i)}\}$, the stress vector is also normal to that surface, and its magnitude is σ_i. In other words, there are no shearing stresses on the surface that is normal to a principal direction.

Without loss of generality, we assume $\sigma_1 \geq \sigma_2 \geq \sigma_3$. Then it can be shown that σ_1 and σ_3 are the maximum and minimum normal stresses, respectively, on all surfaces at a point. The proof is given as follows.

On an arbitrary surface with unit normal vector $\{n\}$, let the stress vector be $\{t\}$. The normal component (projection) of $\{t\}$ on $\{n\}$ is given by

$$\begin{aligned} \sigma_n = \mathbf{t} \cdot \mathbf{n} &= \{t\}^{\mathrm{T}} \{n\} \\ &= ([\sigma]\{n\})^{\mathrm{T}} \{n\} \\ &= \{n\}^{\mathrm{T}} [\sigma]^{\mathrm{T}} \{n\} = \{n\}^{\mathrm{T}} [\sigma] \{n\} \end{aligned} \qquad (2.36)$$

where superscript T indicates the transposed matrix.

Choose a coordinate system so that x, y, and z axes are parallel to the principal directions of stress, respectively. With respect to this coordinate system, all shearing stress components vanish. Then $[\sigma]$ has a simple form as

$$[\sigma] = \begin{bmatrix} \sigma_1 & 0 & 0 \\ 0 & \sigma_2 & 0 \\ 0 & 0 & \sigma_3 \end{bmatrix} \qquad (2.37)$$

Substituting (2.37) into (2.36), we obtain

$$\sigma_n = \sigma_1 n_x^2 + \sigma_2 n_y^2 + \sigma_3 n_z^2 \qquad (2.38)$$

If $\sigma_1 \geq \sigma_2 \geq \sigma_3$, then we have

$$\sigma_1 n_x^2 + \sigma_1 n_y^2 + \sigma_1 n_z^2 \geq \sigma_n \geq \sigma_3 n_x^2 + \sigma_3 n_y^2 + \sigma_3 n_z^2$$

Since $n_x^2 + n_y^2 + n_z^2 = 1$ (**n** is a unit vector), it is obvious that

$$\sigma_1 \geq \sigma_n \geq \sigma_3 \qquad (2.39)$$

Example 2.3 Given the stress matrix

$$[\sigma] = \begin{bmatrix} 2 & 2 & 0 \\ 2 & 2 & 0 \\ 0 & 0 & 1 \end{bmatrix} \qquad \text{MPa}$$

the eigenvalue problem is

$$\begin{vmatrix} 2-\sigma & 2 & 0 \\ 2 & 2-\sigma & 0 \\ 0 & 0 & 1-\sigma \end{vmatrix} = 0$$

which can be expanded into

$$\sigma^3 - 5\sigma^2 + 4\sigma = 0$$

The three roots for the equation above are obtained as

$$\sigma_1 = 4, \qquad \sigma_2 = 1, \qquad \sigma_3 = 0$$

These are the principal stresses.

The unit normal $\mathbf{n}^{(1)}$ corresponding to $\sigma_1 = 4$ MPa can be obtained by substituting this value back into the system of equations (2.33) to obtain

$$-2n_x^{(1)} + 2n_y^{(1)} = 0$$

$$2n_x^{(1)} - 2n_y^{(1)} = 0$$

$$-3n_z^{(1)} = 0$$

Note that these three equations are not independent. Thus, only two equations are available to determine the solution. Since there are three unknowns, two equations can determine only up to the ratios among the three quantities $n_x^{(1)}$, $n_y^{(1)}$, and $n_z^{(1)}$. However, we note that $\mathbf{n}^{(1)}$ is a unit vector, i.e.,

$$(n_x^{(1)})^2 + (n_y^{(1)})^2 + (n_z^{(1)})^2 = 1$$

With this additional equation, the solution is obtained uniquely as

$$n_x^{(1)} = \frac{1}{\sqrt{2}}, \qquad n_y^{(1)} = \frac{1}{\sqrt{2}}, \qquad n_z^{(1)} = 0$$

Following similar manipulations, the unit vectors $\mathbf{n}^{(2)}$ and $\mathbf{n}^{(3)}$ corresponding to σ_2 and σ_3, respectively, can be determined. We have

$$n_x^{(2)} = 0, \qquad n_y^{(2)} = 0, \qquad n_z^{(2)} = 1$$

and

$$n_x^{(3)} = \frac{1}{\sqrt{2}}, \qquad n_y^{(3)} = -\frac{1}{\sqrt{2}}, \qquad n_z^{(3)} = 0$$

It is easy to verify that these three eigenvectors are mutually orthogonal.

2.6 SHEAR STRESS

The stress vector $\mathbf{t}$ can be decomposed into a normal vector $\sigma_n \mathbf{n}$ and a tangential vector τ that is lying on the surface with the unit normal $\mathbf{n}$, i.e.,

$$\mathbf{t} = \sigma_n \mathbf{n} + \tau \tag{2.40}$$

Thus,

$$\tau = \mathbf{t} - \sigma_n \mathbf{n} \tag{2.41}$$

Denoting the magnitude of the shear stress vector by τ, we have

$$\tau^2 = ||\mathbf{t}||^2 - \sigma_n^2 = (t_x^2 + t_y^2 + t_z^2) - \sigma_n^2 \tag{2.42}$$

Let us choose the coordinate system (x, y, z) to be parallel to the principal directions with corresponding principal stresses σ_1, σ_2, and σ_3, respectively. With respect to this coordinate system, the stress matrix $[\sigma]$ assumes the diagonal form as in (2.37). From (2.29) we obtain

$$t_x = \sigma_1 n_x$$

$$t_y = \sigma_2 n_y \tag{2.43}$$

$$t_z = \sigma_3 n_z$$

Using (2.43), the magnitude of the stress vector can be written as

$$||\mathbf{t}||^2 = t_x^2 + t_y^2 + t_z^2$$

$$= (\sigma_1 n_x)^2 + (\sigma_2 n_y)^2 + (\sigma_3 n_z)^2 \tag{2.44}$$

Substituting (2.44) and (2.38) into (2.42), we obtain

$$\tau^2 = (\sigma_1 n_x)^2 + (\sigma_2 n_y)^2 + (\sigma_3 n_z)^2 - (\sigma_1 n_x^2 + \sigma_2 n_y^2 + \sigma_3 n_z^2)^2$$

$$= n_x^2(1 - n_x^2)\sigma_1^2 + n_y^2(1 - n_y^2)\sigma_2^2 + n_z^2(1 - n_z^2)\sigma_3^2$$

$$- 2\sigma_1\sigma_2 n_x^2 n_y^2 - 2\sigma_2\sigma_3 n_y^2 n_z^2 - 2\sigma_1\sigma_3 n_x^2 n_z^2 \tag{2.45}$$

Equation (2.45) can be further simplified by using the relation $1 - n_x^2 = n_y^2 + n_z^2$. We obtain

$$\tau^2 = n_x^2 n_y^2 (\sigma_1 - \sigma_2)^2 + n_y^2 n_z^2 (\sigma_2 - \sigma_3)^2$$

$$+ n_z^2 n_x^2 (\sigma_3 - \sigma_1)^2 \tag{2.46}$$

Consider all surfaces that contain the y-axis, namely surfaces with unit normal vector perpendicular to the y-axis. For any of these surfaces, we have

$$n_x \neq 0, \qquad n_y = 0, \qquad n_z \neq 0 \tag{2.47}$$

From (2.46), the magnitude of the shear stress is

$$\tau^2 = n_z^2 n_x^2 (\sigma_3 - \sigma_1)^2$$

$$= (1 - n_x^2)n_x^2(\sigma_3 - \sigma_1)^2 \tag{2.48}$$

In deriving (2.48), the equation $n_x^2 + n_z^2 = 1$ has been used.

The extremum of $|\tau|$ occurs at

$$\frac{\partial(\tau^2)}{\partial n_x} = 0 = (2n_x - 4n_x^3)(\sigma_3 - \sigma_1)^2 \tag{2.49}$$

This yields the solutions

$$n_x = 0 \quad \text{and} \quad n_x = \pm\frac{1}{\sqrt{2}} \tag{2.50}$$

It can easily be shown that $n_x = 0$ leads to the minimum value of τ^2 and $n_x = \pm 1/\sqrt{2}$ yields the maximum shear stress.

Since $n_y = 0$, and $n_x^2 + n_z^2 = 1$, the solution $n_x = \pm 1/\sqrt{2}$ gives $n_z = \pm 1/\sqrt{2}$. These represent two surfaces making, respectively, $+45°$ and $-45°$ with respect to the x-axis.

Substituting $n_x = \pm 1/\sqrt{2}$ into (2.48), the maximum shear stress is obtained as

$$\tau_{max}^2 = \tfrac{1}{4}(\sigma_3 - \sigma_1)^2 \quad \text{or} \quad |\tau_{max}| = \tfrac{1}{2}|\sigma_3 - \sigma_1| \tag{2.51}$$

Similar considerations of surfaces containing the x and z axes, respectively, yield

$$|\tau_{max}| = \tfrac{1}{2}|\sigma_2 - \sigma_3| \tag{2.52}$$

$$|\tau_{max}| = \tfrac{1}{2}|\sigma_1 - \sigma_2| \tag{2.53}$$

Among these three shear stresses given by (2.51)–(2.53), the one given by (2.51) is the true maximum shear stress if we assume, with no loss of generality,

$$\sigma_1 \geq \sigma_2 \geq \sigma_3$$

2.7 REVISIT OF TRANSFORMATION OF STRESS

Consider a state of plane stress, i.e., $\sigma_{zz} = \tau_{xz} = \tau_{yz} = 0$ and $\sigma_{xx} \neq 0$, $\sigma_{yy} \neq 0$, and $\tau_{xy} \neq 0$. Let $x'-y'$ be coordinate axes obtained by rotating $x-y$ axes a θ angle in the counterclockwise direction (see Fig. 2.12).

Consider the surface perpendicular to the x'-axis. Using (2.28), the stress vector **t** acting on this surface is given by

$$\left\{ \begin{array}{c} t_x \\ t_y \end{array} \right\} = \left[\begin{array}{cc} \sigma_{xx} & \tau_{xy} \\ \tau_{xy} & \sigma_{yy} \end{array} \right] \left\{ \begin{array}{c} n_x \\ n_y \end{array} \right\} \tag{2.54}$$

where $\mathbf{n} = (n_x, n_y)$ is the unit vector parallel to the x'-axis.

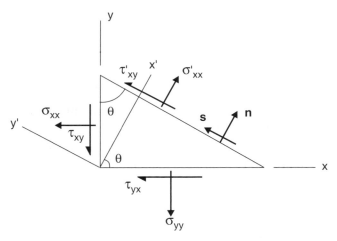

Figure 2.12 Stress components in the $x-y$ and $x'-y'$ coordinates.

Let σ'_{xx}, σ'_{yy}, and τ'_{xy} be the stress components in reference to the $x'-y'$ coordinates. Noting that $\sigma'_{xx} = \sigma_n$, we have

$$\sigma'_{xx} = \sigma_n = \mathbf{t} \cdot \mathbf{n} = t_x n_x + t_y n_y \tag{2.55}$$

Substituting the following relations

$$t_x = \sigma_{xx} n_x + \tau_{xy} n_y$$

$$t_y = \tau_{xy} n_x + \sigma_{yy} n_y$$

into (2.55) yields

$$\sigma'_{xx} = \sigma_{xx} n_x^2 + \tau_{xy} n_x n_y + \tau_{xy} n_x n_y + \sigma_{yy} n_y^2$$

$$= \sigma_{xx} n_x^2 + 2 n_x n_y \tau_{xy} + \sigma_{yy} n_y^2 \tag{2.56}$$

Noting $n_x = \cos\theta$, and $n_y = \sin\theta$, we rewrite (2.56) in the form

$$\sigma'_{xx} = \sigma_{xx} \cos^2\theta + \sigma_{yy} \sin^2\theta + \tau_{xy} \sin 2\theta \tag{2.57}$$

Further use of

$$\cos^2\theta = \tfrac{1}{2}(1 + \cos 2\theta), \qquad \sin^2\theta = \tfrac{1}{2}(1 - \cos 2\theta)$$

in (2.57) leads to

$$\sigma'_{xx} = \tfrac{1}{2}(\sigma_{xx} + \sigma_{yy}) + \tfrac{1}{2}(\sigma_{xx} - \sigma_{yy}) \cos 2\theta + \tau_{xy} \sin 2\theta \tag{2.58}$$

The shear stress component τ'_{xy} can be regarded as the tangential component of the stress vector, i.e.,

$$\tau'_{xy} = \mathbf{t} \cdot \mathbf{s} = t_x s_x + t_y s_y$$

where $\mathbf{s}$ is the unit vector parallel to the y'-axis (or parallel to the surface of interest; see Fig. 2.11), and its two components are

$$s_x = -\sin\theta, \qquad s_y = \cos\theta$$

Thus,

$$\tau'_{xy} = (\sigma_{xx} n_x + \tau_{xy} n_y) s_x + (\tau_{xy} n_x + \sigma_{yy} n_y) s_y$$

$$= -\sigma_{xx} \cos\theta \sin\theta + \tau_{xy}(\cos^2\theta - \sin^2\theta) + \sigma_{yy} \cos\theta \sin\theta$$

$$= \tfrac{1}{2}(\sigma_{yy} - \sigma_{xx}) \sin 2\theta + \tau_{xy} \cos 2\theta \tag{2.59}$$

The transformation for σ'_{yy} is obtained by noting that σ'_{yy} is equal to σ'_{xx} if θ is replaced by $\theta + \pi/2$. From (2.57) we have

$$\sigma'_{yy} = \sigma_{xx} \sin^2 \theta + \sigma_{yy} \cos^2 \theta - \tau_{xy} \sin 2\theta \qquad (2.60)$$

Equations (2.58)–(2.60) are the 2-D stress **coordinate transformation** formulas which can be expressed in matrix form as

$$\left\{ \begin{array}{c} \sigma'_{xx} \\ \sigma'_{yy} \\ \tau'_{xy} \end{array} \right\} = \left[\begin{array}{ccc} \cos^2 \theta & \sin^2 \theta & \sin 2\theta \\ \sin^2 \theta & \cos^2 \theta & -\sin 2\theta \\ -\frac{1}{2}\sin 2\theta & \frac{1}{2}\sin 2\theta & \cos 2\theta \end{array} \right] \left\{ \begin{array}{c} \sigma_{xx} \\ \sigma_{yy} \\ \tau_{xy} \end{array} \right\} \qquad (2.61)$$

where σ'_{xx}, σ'_{yy}, and τ'_{xy} are stress components referring to the x'–y' coordinate system.

2.8 LINEAR STRESS-STRAIN RELATIONS

We have introduced six independent stress components (σ_{xx}, σ_{yy}, σ_{zz}, τ_{yz}, τ_{xz}, τ_{xy}) to describe the load carried by a three-dimensional solid at a point, and six independent strain components (ε_{xx}, ε_{yy}, ε_{zz}, γ_{yz}, γ_{xz}, γ_{xy}) to describe the deformation at a point. Since, in general, deformations are produced by loads, strain components are related to stress components. These stress-strain relations are used to characterize the stiffness of a material. One of the common ways to express the stress-strain relations for a material is to write the stress components as functions of the strain components, or vice versa. If the expressions of these functions of a material do not depend on the Cartesian coordinate systems chosen, this material is **isotropic**. This is equivalent to saying that the stiffness property of the material does not depend on direction. Otherwise, the material is called **anisotropic**.

The subject of developing stress-strain relations (more generally referred to as **constitutive models**) is an important area in mechanics of solids. For the interest of aircraft structures, we focus our attention on constitutive models commonly used to describe lightweight metals and fiber-reinforced composites for which linearly elastic behavior dominates in the range of small strains and loading rate effects are negligible in most applications.

For isotropic materials, the simple tension (uniaxial stress) test is often performed to generate a stress-strain curve as shown in Fig. 2.13. For most structural materials, such uniaxial stress-strain curves almost always consist of an initial linear portion and a nonlinear portion beyond a stress level. Of interest is the yield stress σ_Y beyond which a permanent strain is produced upon unloading, as illustrated in Fig. 2.13. Note that the unloading path is linear and parallel to the initial linear portion. The determination of the yield

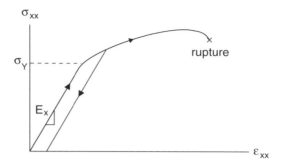

Figure 2.13   Stress-strain curve.

stress is not as easy as the ultimate stress because it does not coincide with the end of the linear portion of the stress-strain curve. A customary practice is to take the stress as yield stress that produces a 0.002 permanent strain upon complete unloading.

2.8.1   Strains Induced by Normal Stress

Imagine that a thin cylindrical element (a rod) is taken out of a generally anisotropic solid in the x-direction. Assume that in this rod, only $\sigma_{xx} \neq 0$, and all the other five stress components are absent (uniaxial stress). If stress σ_{xx} is applied gradually, strains are produced.

Let us consider for now only the normal strain component ε_{xx}. A stress-strain curve as shown in Fig. 2.13 can be obtained. For the linear portion, we write

$$\sigma_{xx} = E_x \varepsilon_{xx} \tag{2.62}$$

in which the constant E_x is called the **Young's modulus** in the x-direction of the solid.

In the case of uniaxial stress loading described above, lateral strains ε_{yy} and ε_{zz} are also present and are related to ε_{xx} as

$$\varepsilon_{yy} = -\nu_{xy}\varepsilon_{xx} \quad \text{or} \quad \nu_{xy} = -\frac{\varepsilon_{yy}}{\varepsilon_{xx}} \tag{2.63}$$

$$\varepsilon_{zz} = -\nu_{xz}\varepsilon_{xx} \quad \text{or} \quad \nu_{xz} = -\frac{\varepsilon_{zz}}{\varepsilon_{xx}} \tag{2.64}$$

where ν_{xy} and ν_{xz} are **Poisson's ratios**. The first subscript indicates the loading direction, and the second subscript indicates the direction of lateral contraction. Using (2.62), (2.63) and (2.64) can be rewritten as

$$\varepsilon_{yy} = -\frac{\nu_{xy}}{E_x}\sigma_{xx} \tag{2.65}$$

$$\varepsilon_{zz} = -\frac{\nu_{xz}}{E_x}\sigma_{xx} \tag{2.66}$$

Similar Young's moduli E_y and E_z in the y and z directions, respectively, are introduced in the uniaxial stress-strain relations as

$$\sigma_{yy} = E_y\varepsilon_{yy} \tag{2.67}$$

and

$$\sigma_{zz} = E_z\varepsilon_{zz} \tag{2.68}$$

respectively. The corresponding Poisson's ratios are introduced in the following relations:

$$\begin{aligned}\varepsilon_{xx} &= -\nu_{yx}\varepsilon_{yy} = -\frac{\nu_{yx}}{E_y}\sigma_{yy}\\[2mm]\varepsilon_{zz} &= -\nu_{yz}\varepsilon_{yy} = -\frac{\nu_{yz}}{E_y}\sigma_{yy}\end{aligned} \tag{2.69}$$

and

$$\begin{aligned}\varepsilon_{xx} &= -\nu_{zx}\varepsilon_{zz} = -\frac{\nu_{zx}}{E_z}\sigma_{zz}\\[2mm]\varepsilon_{yy} &= -\nu_{zy}\varepsilon_{zz} = -\frac{\nu_{zy}}{E_z}\sigma_{zz}\end{aligned} \tag{2.70}$$

It should be noted that in a uniaxially stressed anisotropic body, say $\sigma_{xx} \neq 0$ and other $\sigma_{ij} = 0$, shear strains γ_{yz}, γ_{xz}, and γ_{xy} may be induced in addition to the normal strains. In the most general case, uniaxial stress σ_{xx} may produce six strain components,

$$\begin{aligned}\varepsilon_{xx} &= \frac{1}{E_x}\sigma_{xx}\\[2mm]\varepsilon_{yy} &= -\frac{\nu_{xy}}{E_x}\sigma_{xx}\\[2mm]\varepsilon_{zz} &= -\frac{\nu_{xz}}{E_x}\sigma_{xx}\\[2mm]\gamma_{yz} &= \frac{\eta_{x,yz}}{E_x}\sigma_{xx}\end{aligned} \tag{2.71}$$

$$\gamma_{xz} = \frac{\eta_{x,xz}}{E_x}\sigma_{xx}$$

$$\gamma_{xy} = \frac{\eta_{x,xy}}{E_x}\sigma_{xx}$$

where the coefficients η's serve a similar purpose as Poisson's ratios; they are the ratios of the induced shear strains and the normal strain ε_{xx} produced by uniaxial stress σ_{xx}, i.e.,

$$\eta_{x,yz} = \frac{\gamma_{yz}}{\varepsilon_{xx}},$$

$$\eta_{x,xz} = \frac{\gamma_{xz}}{\varepsilon_{xx}},$$

$$\eta_{x,xy} = \frac{\gamma_{xy}}{\varepsilon_{xx}}$$

Again, the first subscript in η indicates the loading direction, and the second set of subscripts indicates the plane of induced shear strain.

In a similar manner, the strains produced by uniaxial stresses σ_{yy} and σ_{zz} are

$$\varepsilon_{xx} = -\frac{\nu_{yx}}{E_y}\sigma_{yy}$$

$$\varepsilon_{yy} = \frac{1}{E_y}\sigma_{yy}$$

$$\varepsilon_{zz} = -\frac{\nu_{yz}}{E_y}\sigma_{yy}$$

$$\gamma_{yz} = \frac{\eta_{y,yz}}{E_y}\sigma_{yy} \tag{2.72}$$

$$\gamma_{xz} = \frac{\eta_{y,xz}}{E_y}\sigma_{yy}$$

$$\gamma_{xy} = \frac{\eta_{y,xy}}{E_y}\sigma_{yy}$$

and

$$\varepsilon_{xx} = -\frac{\nu_{zx}}{E_z}\sigma_{zz}$$

$$\varepsilon_{yy} = -\frac{\nu_{zy}}{E_z}\sigma_{zz}$$

$$\varepsilon_{zz} = \frac{1}{E_z}\sigma_{zz}$$

$$\gamma_{yz} = \frac{\eta_{z,yz}}{E_z}\sigma_{zz}$$

$$\gamma_{xz} = \frac{\eta_{z,xz}}{E_z}\sigma_{zz} \qquad (2.73)$$

$$\gamma_{xy} = \frac{\eta_{z,xy}}{E_z}\sigma_{zz}$$

respectively. If all three normal stress components are present, then the total strains are the sums of the corresponding strains given by (2.71)–(2.73).

2.8.2 Strains Induced by Shear Stress

Consider a state of simple shear with $\tau_{xy} \neq 0$ and all other stress components are vanishing. In the most general solid, all strain components may be induced by τ_{xy}. Consider the shear strain γ_{xy} induced by τ_{xy}. In the linear range of stress-strain relations, we have

$$\gamma_{xy} = \frac{1}{G_{xy}}\tau_{xy} \qquad (2.74)$$

where G_{xy} is the **shear modulus** in the $x-y$ plane. Other strains induced by τ_{xy} can be written as

$$\varepsilon_{xx} = \eta_{xy,x}\gamma_{xy} = \frac{\eta_{xy,x}}{G_{xy}}\tau_{xy}$$

$$\varepsilon_{yy} = \eta_{xy,y}\gamma_{xy} = \frac{\eta_{xy,y}}{G_{xy}}\tau_{xy}$$

$$\varepsilon_{zz} = \eta_{xy,z}\gamma_{xy} = \frac{\eta_{xy,z}}{G_{xy}}\tau_{xy} \qquad (2.75)$$

$$\gamma_{yz} = \mu_{xy,yz}\gamma_{xy} = \frac{\mu_{xy,yz}}{G_{xy}}\tau_{xy}$$

$$\gamma_{xz} = \mu_{xy,xz}\gamma_{xy} = \frac{\mu_{xy,xz}}{G_{xy}}\tau_{xy}$$

in which $\mu_{xy,yz}$ and $\mu_{xy,xz}$ are introduced to represent the interactions among the shear strains; and $\eta_{xy,x}$, $\eta_{xy,y}$, and $\eta_{xy,z}$ are the interactions between the shear strain γ_{xy} and the normal strains ε_{xx}, ε_{yy}, and ε_{zz}, respectively.

For simple shear in the $y-z$ and $x-z$ planes, we have

$$\gamma_{yz} = \frac{1}{G_{yz}}\tau_{yz} \tag{2.76}$$

and

$$\gamma_{xz} = \frac{1}{G_{xz}}\tau_{xz} \tag{2.77}$$

respectively. Other strains produced by τ_{yz} and τ_{xz}, can be expressed in a form similar to (2.75).

If the material is isotropic, i.e., its mechanical properties are not direction dependent, all the η coefficients vanish and

$$E_x = E_y = E_z = E$$
$$\nu_{xy} = \nu_{yx} = \nu_{xz} = \nu_{zx} = \nu_{yz} = \nu_{zy} = \nu \tag{2.78}$$
$$G_{xy} = G_{xz} = G_{yz} = G$$

Thus, in an isotropic solid, a normal stress does not produce shear strains, and a shear stress does not produce normal strains.

2.8.3   Three-Dimensional Stress-Strain Relations

The discussion in Sections 2.8.1 and 2.8.2 indicates that, in the most general case, the application of a single stress component can possibly produce all six strain components. In the linear range of stress-strain relations, we can write the strains produced by all six stress components by using the principle of superposition:

$$\varepsilon_{xx} = a_{11}\sigma_{xx} + a_{12}\sigma_{yy} + a_{13}\sigma_{zz} + a_{14}\tau_{yz} + a_{15}\tau_{xz} + a_{16}\tau_{xy}$$
$$\varepsilon_{yy} = a_{21}\sigma_{xx} + a_{22}\sigma_{yy} + a_{23}\sigma_{zz} + a_{24}\tau_{yz} + a_{25}\tau_{xz} + a_{26}\tau_{xy}$$
$$\varepsilon_{zz} = a_{31}\sigma_{xx} + a_{32}\sigma_{yy} + a_{33}\sigma_{zz} + a_{34}\tau_{yz} + a_{35}\tau_{xz} + a_{36}\tau_{xy}$$
$$\gamma_{yz} = a_{41}\sigma_{xx} + a_{42}\sigma_{yy} + a_{43}\sigma_{zz} + a_{44}\tau_{yz} + a_{45}\tau_{xz} + a_{46}\tau_{xy} \tag{2.79}$$
$$\gamma_{xz} = a_{51}\sigma_{xx} + a_{52}\sigma_{yy} + a_{53}\sigma_{zz} + a_{54}\tau_{yz} + a_{55}\tau_{xz} + a_{56}\tau_{xy}$$
$$\gamma_{xy} = a_{61}\sigma_{xx} + a_{62}\sigma_{yy} + a_{63}\sigma_{zz} + a_{64}\tau_{yz} + a_{65}\tau_{xz} + a_{66}\tau_{xy}$$

where a_{ij} $(i, j = 1 - 6)$ are **elastic compliances**. Comparing (2.79) with (2.71)–(2.75) together with two similar equations for τ_{yz} and τ_{xz}, we can easily relate the elastic compliances to the engineering moduli E_x, E_y, ..., ν_{xy}, ..., G_{xy}, ..., $\eta_{x,yz}$, ... We have

$$[a_{ij}] = \begin{bmatrix} \dfrac{1}{E_x} & -\dfrac{\nu_{yx}}{E_y} & -\dfrac{\nu_{zx}}{E_z} & \dfrac{\eta_{yz,x}}{G_{yz}} & \dfrac{\eta_{xz,x}}{G_{xz}} & \dfrac{\eta_{xy,x}}{G_{xy}} \\[2ex] -\dfrac{\nu_{xy}}{E_x} & \dfrac{1}{E_y} & -\dfrac{\nu_{zy}}{E_z} & \dfrac{\eta_{yz,y}}{G_{yz}} & \dfrac{\eta_{xz,y}}{G_{xz}} & \dfrac{\eta_{xy,y}}{G_{xy}} \\[2ex] -\dfrac{\nu_{xz}}{E_x} & -\dfrac{\nu_{yz}}{E_y} & \dfrac{1}{E_z} & \dfrac{\eta_{yz,z}}{G_{yz}} & \dfrac{\eta_{xz,z}}{G_{xz}} & \dfrac{\eta_{xy,z}}{G_{xy}} \\[2ex] \dfrac{\eta_{x,yz}}{E_x} & \dfrac{\eta_{y,yz}}{E_y} & \dfrac{\eta_{z,yz}}{E_z} & \dfrac{1}{G_{yz}} & \dfrac{\mu_{xz,yz}}{G_{xz}} & \dfrac{\mu_{xy,yz}}{G_{xy}} \\[2ex] \dfrac{\eta_{x,xz}}{E_x} & \dfrac{\eta_{y,xz}}{E_y} & \dfrac{\eta_{z,xz}}{E_y} & \dfrac{\mu_{yz,xz}}{G_{yz}} & \dfrac{1}{G_{xz}} & \dfrac{\mu_{xy,xz}}{G_{xy}} \\[2ex] \dfrac{\eta_{x,xy}}{E_x} & \dfrac{\eta_{y,xy}}{E_y} & \dfrac{\eta_{z,xy}}{E_z} & \dfrac{\mu_{yz,xy}}{G_{yz}} & \dfrac{\mu_{xz,xy}}{G_{xz}} & \dfrac{1}{G_{xy}} \end{bmatrix} \qquad (2.80)$$

In matrix notation, (2.79) can be expressed as

$$\begin{Bmatrix} \varepsilon_{xx} \\ \varepsilon_{yy} \\ \varepsilon_{zz} \\ \gamma_{yz} \\ \gamma_{xz} \\ \gamma_{xy} \end{Bmatrix} = [a_{ij}] \begin{Bmatrix} \sigma_{xx} \\ \sigma_{yy} \\ \sigma_{zz} \\ \tau_{yz} \\ \tau_{xz} \\ \tau_{xy} \end{Bmatrix} \qquad (2.81)$$

or symbolically as

$$\{\varepsilon\} = [\,a\,]\{\sigma\} \qquad (2.82)$$

The inverse relations of (2.82) are given as

$$\{\sigma\} = [\,c\,]\{\varepsilon\} \qquad (2.83)$$

where

$$[\,c\,] = [\,a\,]^{-1}$$

The relations given by (2.82) and (2.83) are generally called **Hooke's law**. The elements c_{ij} in $[\,c\,]$ are called **elastic constants**, and E_x, E_y, ..., ν_{xy}, ..., G_{xy}, ... are called **engineering moduli**.

   In general, it is easier to measure elastic compliances a_{ij} (and thus the engineering moduli) than elastic constants c_{ij} because a_{ij} can be measured using simple tension and simple shear tests. For instance, under simple tension in the y-direction, the resulting strain components are equal to $a_{i2}\sigma_{yy}$. From the measured strains, the second column a_{i2} of the elastic compliance matrix can thus be obtained. Theoretically, elastic constants c_{ij} may be determined with a similar approach by conducting experiments in which only one strain component is present in each experiment. However, these experiments are much more difficult than uniaxial stress tests to realize in the laboratory.

From the elastic strain energy consideration to be discussed in Section 2.9, it can be shown that [a], and thus [c] also, is a symmetric matrix, i.e.,

$$a_{ij} = a_{ji}, \qquad i, j = 1, 2, \ldots, 6$$

Thus, there are only 21 independent elastic compliances or, equivalently, 21 elastic constants. This is the maximum number of independent elastic constants that linear elastic materials can have. Most materials possess certain **elastic symmetries** that would reduce the number of independent elastic constants. The following are two material groups that are of practical interest.

Orthotropic Materials   Unidirectional fiber composites can be regarded as **orthotropic materials** which possess three mutually orthogonal planes of symmetry. The directions perpendicular to these planes are called the material principal directions. If the coordinates x, y, z are set up such that they are parallel to the material principal directions, respectively, then the elastic compliance matrix reduces to

$$[a_{ij}] = \begin{bmatrix} a_{11} & a_{12} & a_{13} & 0 & 0 & 0 \\ a_{21} & a_{22} & a_{23} & 0 & 0 & 0 \\ a_{31} & a_{32} & a_{33} & 0 & 0 & 0 \\ 0 & 0 & 0 & a_{44} & 0 & 0 \\ 0 & 0 & 0 & 0 & a_{55} & 0 \\ 0 & 0 & 0 & 0 & 0 & a_{66} \end{bmatrix} \qquad (2.84)$$

The elastic compliances a_{ij} are related to the engineering moduli as

$$a_{11} = \frac{1}{E_x}, \qquad a_{12} = -\frac{v_{yx}}{E_y}, \qquad a_{13} = -\frac{v_{zx}}{E_z}$$

$$a_{21} = -\frac{v_{xy}}{E_x}, \qquad a_{22} = \frac{1}{E_y}, \qquad a_{23} = -\frac{v_{zy}}{E_z}$$

$$a_{31} = -\frac{v_{xz}}{E_x}, \qquad a_{32} = -\frac{v_{yz}}{E_y}, \qquad a_{33} = \frac{1}{E_z} \qquad (2.85)$$

$$a_{44} = \frac{1}{G_{yz}}, \qquad a_{55} = \frac{1}{G_{xz}}, \qquad a_{66} = \frac{1}{G_{xy}}$$

Since $a_{ij} = a_{ji}$, we have

$$\nu_{yx} = \frac{E_y}{E_x}\nu_{xy}, \qquad \nu_{zx} = \frac{E_z}{E_x}\nu_{xz}, \qquad \nu_{zy} = \frac{E_z}{E_y}\nu_{yz} \qquad (2.86)$$

Thus, there are nine independent elastic constants for orthotropic elastic materials.

Fiber-reinforced composites are regarded as orthotropic solids. It is customary to denote the fiber direction as x_1-axis, and the transverse directions as x_2 and x_3. The elastic moduli are referenced to this particular coordinate system and denoted by E_1, E_2, E_3, ν_{12}, ν_{13}, ν_{23}, G_{23}, G_{13}, and G_{12}.

The following are the in-plane engineering moduli for some polymeric composites.

AS4/3501-6 carbon/epoxy   (AS4 carbon fiber in 3501-6 epoxy):

$$E_1 = 140\,\text{GPa}, \qquad E_2 = 10\,\text{GPa}$$
$$G_{12} = 7.0\,\text{GPa}, \qquad \nu_{12} = 0.3$$

Boron/epoxy:

$$E_1 = 205\,\text{GPa}, \qquad E_2 = 21\,\text{GPa}$$
$$G_{12} = 5.5\,\text{GPa}, \qquad \nu_{12} = 0.17$$

S2-glass/epoxy:

$$E_1 = 43\,\text{GPa}, \qquad E_2 = 9\,\text{GPa}$$
$$G_{12} = 4.5\,\text{GPa}, \qquad \nu_{12} = 0.27$$

Note that the Young's modulus in the fiber direction (E_1) for AS4/3501-6 is 14 times the transverse Young's modulus (E_2).

Isotropic Materials   The elastic properties of isotropic materials are invariant with respect to directions. Thus, isotropy is a special case of orthotropy. By requiring the conditions given by (2.78), we obtain

$$a_{11} = a_{22} = a_{33} = \frac{1}{E}$$

$$a_{12} = a_{13} = a_{23} = -\frac{\nu}{E} \qquad (2.87)$$

$$a_{44} = a_{55} = a_{66} = \frac{1}{G}$$

The corresponding elastic constants c_{ij} can also be expressed in terms of engineering moduli as

$$c_{11} = c_{22} = c_{33} = \lambda + 2G$$
$$c_{12} = c_{13} = c_{23} = \lambda \qquad (2.88)$$
$$c_{44} = c_{55} = c_{66} = G$$

where $\lambda = \nu E / (1 + \nu)(1 - 2\nu)$.

It is evident that the stress-strain relations for isotropic materials can be expressed in terms of the Young's modulus E, Poisson's ratio ν, and shear modulus G. Moreover, it can be shown that these three quantities are related by

$$E = 2(1 + \nu)G \qquad (2.89)$$

Thus, there are only two independent elastic constants for isotropic materials.

In terms of stiffness, aluminum alloys are usually considered isotropic materials. Typical values of their elastic moduli are

$$E = 70 \text{ GPa}, \qquad \nu = 0.33$$

2.9 ELASTIC STRAIN ENERGY

An elastic body can store energy in the form of deformation. This strain energy is *completely* released when loads are removed. Since the strain energy is stored solely in the form of deformation, it can be expressed in terms of strain components or stress components.

Consider an infinitesimal solid element $\Delta x \times \Delta y \times \Delta z$ as shown in Fig. 2.14. On the x-face, only the normal stress σ_{xx} is present. The corresponding normal strain is ε_{xx}. The total force acting on the x-face is $\Delta y \Delta z \sigma_{xx}$ and the elongation of the element in the x-direction is $\Delta x \varepsilon_{xx}$. The work done by σ_{xx} is converted entirely to strain energy. Thus, the strain energy stored in the element is

$$\Delta U = \tfrac{1}{2}(\Delta y \Delta z \sigma_{xx})(\Delta x \varepsilon_{xx}) = \tfrac{1}{2}(\Delta V \sigma_{xx} \varepsilon_{xx}) \qquad (2.90)$$

where ΔV is the volume of the element. The factor $\tfrac{1}{2}$ in (2.90) accounts for the linear stress-strain relations of the elastic solid.

If all three normal stress components are present, then the strain energy stored in the element is

$$\Delta U = \tfrac{1}{2}\Delta V (\sigma_{xx} \varepsilon_{xx} + \sigma_{yy} \varepsilon_{yy} + \sigma_{zz} \varepsilon_{zz}) \qquad (2.91)$$

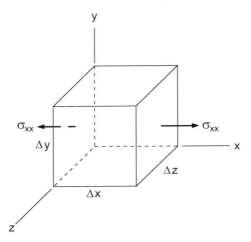

Figure 2.14   Infinitesimal solid element subjected to normal stress σ_{xx}.

If the infinitesimal element is subjected to the shear stress τ_{xz} as shown in Fig. 2.15, the work done (and thus the strain energy) is

$$\Delta U = \tfrac{1}{2}(\Delta x\, \Delta y\, \tau_{xz})(\gamma_{xz}\, \Delta z) \tag{2.92}$$

where the first part on the right-hand side represents the total force acting on the top face, and the second part represents the distance the shear force travels during the deformation. It is noted that the shear forces acting on other faces of the element do not contribute to the work.

The total strain energy produced by all three shear stress components is

$$\Delta U = \tfrac{1}{2}\Delta V (\tau_{xz}\gamma_{xz} + \tau_{xy}\gamma_{xy} + \tau_{yz}\gamma_{yz}) \tag{2.93}$$

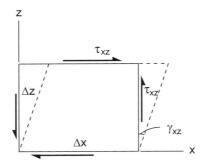

Figure 2.15   Infinitesimal element subjected to shear stress τ_{xz}.

The total strain energy stored in ΔV due to normal and shear stresses is the sum of (2.91) and (2.93). The **strain energy density** W is defined as

$$W = \frac{\Delta U}{\Delta V} = \frac{1}{2}(\sigma_{xx}\varepsilon_{xx} + \sigma_{yy}\varepsilon_{yy} + \sigma_{zz}\varepsilon_{zz}$$
$$+ \tau_{xz}\gamma_{xz} + \tau_{xy}\gamma_{xy} + \tau_{yz}\gamma_{yz}) \qquad (2.94)$$

Using the matrix representations of the stress-strain relations (2.82) and (2.83), the strain energy density can be expressed in various forms as

$$W = \tfrac{1}{2}\{\sigma\}^{\mathrm{T}}\{\varepsilon\} = \tfrac{1}{2}\{\varepsilon\}^{\mathrm{T}}\{\sigma\}$$
$$= \tfrac{1}{2}\{\varepsilon\}^{\mathrm{T}}[\,c\,]\{\varepsilon\}$$
$$= \tfrac{1}{2}\{\sigma\}^{\mathrm{T}}[\,a\,]\{\sigma\} \qquad (2.95)$$

Since $W > 0$, the quadratic forms above imply that $[\,c\,]$ and $[\,a\,]$ must be positive definite. Consequently, all the diagonal terms $c_{ii}\,(i = 1 - 6)$ and $a_{ii}\,(i = 1 - 6)$ must be positive. For orthotropic materials, this implies that all the major engineering moduli E_x, E_y, E_z, G_{yz}, G_{xz}, and G_{xy} cannot assume negative values.

Strain energy density is a real number. If it is regarded as a 1×1 matrix, then

$$[W]^{\mathrm{T}} = [W]$$

By using (2.95), the relation above leads to

$$\{\sigma\}^{\mathrm{T}}[a]^{\mathrm{T}}\{\sigma\} = \{\sigma\}^{\mathrm{T}}[a]\{\sigma\}$$

and thus, $[a]^{\mathrm{T}} = [a]$, i.e., $a_{ij} = a_{ji}$.

2.10 PLANE ELASTICITY

Many structures under certain types of loading may yield strain and stress fields that are independent of, say, the z-direction and have

$$\varepsilon_{zz} = \gamma_{yz} = \gamma_{xz} = 0 \qquad (2.96)$$

or

$$\sigma_{zz} = \tau_{xz} = \tau_{yz} = 0 \qquad (2.97)$$

over the entire structure. The state of deformation with (2.96) is called a state of **plane strain** parallel to the x–y plane, while that with (2.97) is called a state of **plane stress** parallel to the x–y plane.

From the strain–displacement relations, the conditions of (2.96) can be described in terms of the displacement field as

$$u = u(x, y)$$
$$v = v(x, y) \tag{2.98}$$
$$w = 0$$

where u, v, and w are displacement components in the x, y, and z directions, respectively.

To produce a state of plane strain parallel to the $x-y$ plane, the structure must be uniform in shape in the z-direction, and loading must be independent of the z-axis. An example is a hollow cylinder subjected to uniform internal pressure with both ends constrained to suppress its movement in the z-direction.

The plane stress condition given by (2.97), in general, cannot be exactly realized. It is often used to approximate the state of stress in a thin panel subjected to only in-plane ($x-y$ plane) loading. The corresponding displacement field is given by

$$u = u(x, y)$$
$$v = v(x, y) \tag{2.99}$$
$$w = \varepsilon_{zz} z$$

where normal strain ε_{zz} is independent of the z-axis.

2.10.1 Stress-Strain Relations for Plane Isotropic Solids

Plane strain and plane stress conditions lead to simplification in stress-strain relations. Recall that the 3-D stress-strain relations for isotropic solids can be expressed in the following forms:

$$\{\sigma\} = [c]\{\varepsilon\} \tag{2.100}$$

or

$$\{\varepsilon\} = [a]\{\sigma\} \tag{2.101}$$

where

$$c_{11} = c_{22} = c_{33} = \lambda + 2G = \frac{E(1-v)}{(1+v)(1-2v)}$$

$$c_{12} = c_{21} = c_{13} = c_{31} = c_{23} = c_{32} = \lambda = \frac{Ev}{(1+v)(1-2v)}$$

$$c_{44} = c_{55} = c_{66} = G$$

other $c_{ij} = 0$

$$a_{11} = a_{22} = a_{33} = \frac{1}{E}$$

$$a_{12} = a_{21} = a_{13} = a_{31} = a_{23} = a_{32} = -\frac{\nu}{E}$$

$$a_{44} = a_{55} = a_{66} = \frac{1}{G}$$

other $a_{ij} = 0$

Plane Strain   Using the plane strain conditions, (2.100) reduces to

$$\sigma_{xx} = \frac{E}{(1+\nu)(1-2\nu)}[(1-\nu)\varepsilon_{xx} + \nu\varepsilon_{yy}]$$

$$\sigma_{yy} = \frac{E}{(1+\nu)(1-2\nu)}[\nu\varepsilon_{xx} + (1-\nu)\varepsilon_{yy}]$$

$$\sigma_{zz} = \frac{\nu E}{(1+\nu)(1-2\nu)}[\varepsilon_{xx} + \varepsilon_{yy}] \qquad (2.102)$$

$$\tau_{yz} = 0$$

$$\tau_{xz} = 0$$

$$\tau_{xy} = G\gamma_{xy}$$

From the first three equations of (2.102), it is easy to see that

$$\sigma_{zz} = \nu(\sigma_{xx} + \sigma_{yy})$$

Thus, σ_{zz} is a dependent quantity. The first, second, and last equations in (2.102) are usually considered the stress-strain relations for a state of plane strain. Inverting these equations, we obtain

$$\varepsilon_{xx} = \frac{1+\nu}{E}[(1-\nu)\sigma_{xx} - \nu\sigma_{yy}]$$

$$\varepsilon_{yy} = \frac{1+\nu}{E}[-\nu\sigma_{xx} + (1-\nu)\sigma_{yy}] \qquad (2.103)$$

$$\gamma_{xy} = \frac{1}{G}\tau_{xy}$$

Plane Stress   The reduced stress-strain relations for a state of plane stress are readily derived from the 3-D relations given by (2.101). We have

$$\varepsilon_{xx} = \frac{1}{E}(\sigma_{xx} - \nu\sigma_{yy})$$

$$\varepsilon_{yy} = \frac{1}{E}(-\nu\sigma_{xx} + \sigma_{yy})$$

$$\varepsilon_{zz} = -\frac{\nu}{E}(\sigma_{xx} + \sigma_{yy}) \qquad (2.104)$$

$$\gamma_{yz} = 0$$

$$\gamma_{xz} = 0$$

$$\gamma_{xy} = \frac{1}{G}\tau_{xy}$$

Again, from (2.104) we note that

$$\varepsilon_{zz} = -\frac{\nu}{1-\nu}(\varepsilon_{xx} + \varepsilon_{yy}) \qquad (2.105)$$

is a dependent quantity. Inverting the first two and last equations in (2.104), we obtain

$$\sigma_{xx} = \frac{E}{1-\nu^2}(\varepsilon_{xx} + \nu\varepsilon_{yy})$$

$$\sigma_{yy} = \frac{E}{1-\nu^2}(\nu\varepsilon_{xx} + \varepsilon_{yy}) \qquad (2.106)$$

$$\tau_{xy} = G\gamma_{xy}$$

These relations are different from the corresponding plane strain relations given by (2.102). However, the two sets of stress-strain relations can be put in a single expression by introducing the following parameter defined by

$$\text{For plane strain:} \quad \kappa = 3 - 4\nu$$

$$\text{For plane stress:} \quad \kappa = \frac{3-\nu}{1+\nu}$$

In terms of parameter κ, the stress-strain relations for both plane strain and plane stress can be expressed in the form

$$\sigma_{xx} = 2G\left[\varepsilon_{xx} + \frac{3-\kappa}{2\kappa-2}(\varepsilon_{xx} + \varepsilon_{yy})\right]$$

$$\sigma_{yy} = 2G\left[\varepsilon_{yy} + \frac{3-\kappa}{2\kappa-2}(\varepsilon_{xx} + \varepsilon_{yy})\right] \qquad (2.107)$$

$$\tau_{xy} = G\gamma_{xy}$$

or

$$\varepsilon_{xx} = \frac{1}{2G}\left[\sigma_{xx} - \frac{3-\kappa}{4}(\sigma_{xx}+\sigma_{yy})\right]$$

$$\varepsilon_{yy} = \frac{1}{2G}\left[\sigma_{yy} - \frac{3-\kappa}{4}(\sigma_{xx}+\sigma_{yy})\right] \qquad (2.108)$$

$$\gamma_{xy} = \frac{1}{G}\tau_{xy}$$

2.10.2 Stress-Strain Relations for Orthotropic Solids in Plane Stress

Fiber-reinforced composite materials are modeled as orthotropic solids. Furthermore, they are often used in the form of thin panels for which the plane stress condition prevails.

Let the fiber direction coincide with the x-axis and the panel be parallel to the x–y plane. The stress-strain relations for a composite panel are given by

$$\left\{\begin{array}{c}\varepsilon_{xx}\\\varepsilon_{yy}\\\gamma_{xy}\end{array}\right\} = \begin{bmatrix}\dfrac{1}{E_1} & -\dfrac{\nu_{21}}{E_2} & 0\\[2mm]-\dfrac{\nu_{12}}{E_1} & \dfrac{1}{E_2} & 0\\[2mm]0 & 0 & \dfrac{1}{G_{12}}\end{bmatrix}\left\{\begin{array}{c}\sigma_{xx}\\\sigma_{yy}\\\tau_{xy}\end{array}\right\} \qquad (2.109)$$

where E_1 is the Young's modulus in the fiber direction (the longitudinal modulus), E_2 is the transverse Young's modulus, G_{12} is the shear modulus in the x–y plane, and ν_{12} and ν_{21} are Poisson's ratios. Since the compliance matrix is symmetric, we have

$$\frac{\nu_{21}}{E_2} = \frac{\nu_{12}}{E_1} \qquad (2.110)$$

Thus, there are only four independent elastic moduli.

Inverting relations (2.109), we have

$$\left\{\begin{array}{c}\sigma_{xx}\\\sigma_{yy}\\\tau_{xy}\end{array}\right\} = \begin{bmatrix}\dfrac{E_1}{\Delta} & \dfrac{\nu_{12}E_2}{\Delta} & 0\\[2mm]\dfrac{\nu_{12}E_2}{\Delta} & \dfrac{E_2}{\Delta} & 0\\[2mm]0 & 0 & G_{12}\end{bmatrix}\left\{\begin{array}{c}\varepsilon_{xx}\\\varepsilon_{yy}\\\gamma_{xy}\end{array}\right\} \qquad (2.111)$$

where

$$\Delta = 1 - \nu_{12}\nu_{21}$$

2.10.3 Governing Equations

To solve plane elasticity problems, three sets of conditions must be satisfied, i.e., the equilibrium equations, boundary conditions, and compatibility equations.

Equilibrium Equations   For plane problems, it is easy to show that the 3-D equilibrium equations (2.21)–(2.23) reduce to

$$\frac{\partial \sigma_{xx}}{\partial x} + \frac{\partial \tau_{xy}}{\partial y} = 0$$
$$\frac{\partial \tau_{xy}}{\partial x} + \frac{\partial \sigma_{yy}}{\partial y} = 0 \tag{2.112}$$

Boundary Conditions   For plane problems, the loading stress vector **t** on the boundary is in the x–y plane, i.e., $t_z = 0$. On the boundary contour, the applied traction (stress vector) is given and the stresses must satisfy the following boundary conditions:

$$\left\{ \begin{array}{c} t_x \\ t_y \end{array} \right\} = \left[\begin{array}{cc} \sigma_{xx} & \tau_{xy} \\ \tau_{xy} & \sigma_{yy} \end{array} \right] \left\{ \begin{array}{c} n_x \\ n_y \end{array} \right\} \tag{2.113}$$

where $\mathbf{n} = (n_x, n_y)$ is the unit normal vector to the boundary contour of the plane body.

Compatibility Equation   The three strain components ε_{xx}, ε_{yy}, and γ_{xy} are derived from the two displacement components u and v as

$$\varepsilon_{xx} = \frac{\partial u}{\partial x}, \qquad \varepsilon_{yy} = \frac{\partial v}{\partial y}, \qquad \gamma_{xy} = \frac{\partial u}{\partial y} + \frac{\partial v}{\partial x} \tag{2.114}$$

Using the strain-displacement relations in (2.114), we can derive the following compatibility equation:

$$\frac{\partial^2 \varepsilon_{xx}}{\partial y^2} + \frac{\partial^2 \varepsilon_{yy}}{\partial x^2} = \frac{\partial^2 \gamma_{xy}}{\partial x\, \partial y} \tag{2.115}$$

Thus, the three strain functions ε_{xx}, ε_{yy}, and γ_{xy} cannot be arbitrarily specified; they must satisfy the compatibility equation (2.115). Otherwise, we may not be able to find a unique displacement field.

For isotropic solids, the compatibility equation (2.115) can be written in terms of stresses by using the stress-strain relations and equilibrium equations. We have

$$\nabla^2(\sigma_{xx} + \sigma_{yy}) = 0 \tag{2.116}$$

where

$$\nabla^2 = \frac{\partial^2}{\partial x^2} + \frac{\partial^2}{\partial y^2}$$

In general three-dimensional problems, there are six strain components. Following the procedure used in deriving (2.115), we are able to derive five additional compatibility equations that involve the strain components. In the case of plane strain problems, among the six strain components, only the in-plane strains ε_{xx}, ε_{yy}, and γ_{xy} are not vanishing and, thus, only the compatibility equation (2.115) is not trivial.

For plane stress, normal strain ε_{zz} is present in addition to the three in-plane strain components. If we assume that all stress components (and thus all strain components) are independent of the z-axis, then, besides (2.115), there are additional compatibility equations that need to be satisfied. They are

$$\frac{\partial^2 \varepsilon_{zz}}{\partial x^2} = 0, \qquad \frac{\partial^2 \varepsilon_{zz}}{\partial y^2} = 0, \qquad \frac{\partial^2 \varepsilon_{zz}}{\partial x \, \partial y} = 0 \tag{2.117}$$

The equations above indicate that ε_{zz} must be a linear function of x and y. In general, this contradicts the result of relation (2.105), in which the two in-plane normal strains may not be linear in x and y. In other words, solutions obtained with the plane stress governing equations (2.112), (2.113), and (2.115) do not satisfy all compatibility equations and are, in general, only approximate solutions. However, these solutions are very good for thin plates under in-plane loads. More details on this topic can be found in many books on the theory of elasticity.[1]

The plane stress formulation can be an exact formulation for some special states of stress. For example, a uniform state of stress in plates produced by loads that are applied uniformly over the thickness along the edge of the plate satisfies all the compatibility equations. Hence, such a state of stress is plane stress rather than plane strain even if plate thickness is large.

---

[1] See, for example, S. P. Timoshenko and J. N. Goodier, *Theory of Elasticity*, 3rd ed., McGraw-Hill, New York, 1970, p. 274.

2.10.4 Solution by Airy Stress Function for Plane Isotropic Solids

Consider the possibility of the existence of a function $\phi(x, y)$ such that

$$\sigma_{xx} = \frac{\partial^2 \phi}{\partial y^2}$$

$$\sigma_{yy} = \frac{\partial^2 \phi}{\partial x^2} \qquad (2.118)$$

$$\tau_{xy} = -\frac{\partial^2 \phi}{\partial x\, \partial y}$$

With the relations given by (2.118), it is easy to verify that the equilibrium equations (2.112) are satisfied automatically. Substitution of (2.118) into (2.116) yields

$$\nabla^2 \nabla^2 \phi = 0$$

or, explicitly,

$$\frac{\partial^4 \phi}{\partial x^4} + 2\frac{\partial^4 \phi}{\partial x^2\, \partial y^2} + \frac{\partial^4 \phi}{\partial y^4} = 0 \qquad (2.119)$$

The above is the compatibility equation in terms of ϕ which is called the **Airy stress function**. In this form, the solution to a plane elasticity problem is reduced to solving (2.119) for ϕ from which stresses are derived from (2.118). These stresses are required to satisfy the boundary conditions. Note that, in using the Airy stress function, one need not worry about equilibrium equations since they are automatically satisfied. The solution procedure involves solving the partial differential equation (2.119) and satisfying the boundary conditions (2.113).

As an example, consider a beam of rectangular cross-section subjected to pure bending as shown in Fig. 2.16. This can be considered as a 2-D plane

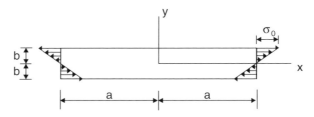

Figure 2.16   Beam under pure bending.

stress problem. It is easy to verify that the Airy stress function

$$\phi = Cy^3 \tag{2.120}$$

is the solution to the bending problem. To show that this Airy stress function is the solution, we must check whether it satisfies the compatibility equation (2.119) and boundary conditions which state:

$$\sigma_{xx} = \sigma_0 y/b \text{ and } \tau_{xy} = 0 \qquad \text{on the vertical faces at } x = \pm a$$

$$\sigma_{yy} = 0 \text{ and } \tau_{xy} = 0 \qquad \text{on the top and bottom faces at } y = \pm b$$

It is obvious that this Airy stress function satisfies the compatibility equation (2.119). Using (2.118), we obtain the stress components corresponding to the Airy stress function as

$$\sigma_{xx} = 6Cy$$
$$\sigma_{yy} = 0$$
$$\tau_{xy} = 0$$

Thus, if we choose $C = 1/6b$, then the boundary conditions are totally satisfied.

PROBLEMS

2.1 Consider a unit cube of a solid occupying the region

$$0 \le x \le 1, \qquad 0 \le y \le 1, \qquad 0 \le z \le 1$$

After loads are applied, the displacements are given by

$$u = \alpha x$$
$$v = \beta y$$
$$w = 0$$

(a) Sketch the deformed shape for $\alpha = 0.03$, $\beta = -0.01$.

(b) Calculate the six strain components.

(c) Find the volume change ΔV [$\Delta V = V$ (the volume after deformation) $- V_0$ (the original volume)] for this unit cube. Show that $\varepsilon_{xx} + \varepsilon_{yy} + \varepsilon_{zz} \approx \Delta V$.

2.2 Consider the following displacement field:

$$u = \alpha y$$
$$v = -\alpha x$$
$$w = 0$$

Sketch the displaced configuration of a unit cube with the faces orig-inally perpendicular to the axes, respectively. This displacement field does not yield any strains; it only produces a rigid body rotation. Show that the angle of rotation is

$$\frac{1}{2}\left(\frac{\partial v}{\partial x} - \frac{\partial u}{\partial y}\right) = -\alpha$$

2.3 Consider the displacement field in a body

$$u = \quad 0.02x + 0.02y - 0.01z \text{ cm}$$
$$v = \quad\quad\quad\quad 0.01y - 0.02z \text{ cm}$$
$$w = -0.01x \quad\quad\quad + 0.01z \text{ cm}$$

Find the locations of the two points $(0, 0, 0)$ and $(5, 0, 0)$ after de-formation. What is the change of distance between these two points after deformation? Calculate the strain components corresponding to the given displacement field. Use the definition of ε_{xx} to estimate the change of distance between the two points. Compare the two results.

2.4 Consider the problem of simple shear in Example 2.1 and Fig. 2.5. From the deformed shape, find the normal strain for material along the line $\overline{CB}$ by comparing the deformed length $\overline{C'B'}$ and undeformed length $\overline{CB}$.

Set up new coordinates (x', y') so that the x'-axis coincides with $\overline{CB}$, and y' is perpendicular to the x'-axis. The relation between (x, y) and (x', y') is given by

$$x' = x \cos\theta + y \sin\theta$$
$$y' = -x \sin\theta + y \cos\theta$$

where $\theta = 45°$ is the angle between x' and the x-axis.

Write the displacements u' and v' in the x' and y' directions, respec-tively, in terms of the new coordinates x' and y'. The relation between (u', v') and (u, v) is the same as between (x', y') and (x, y). Then calculate the strains using u' and v', i.e.,

$$\varepsilon'_{xx} = \frac{\partial u'}{\partial x'}$$

$$\varepsilon'_{yy} = \frac{\partial v'}{\partial y'}$$

$$\gamma'_{xy} = \frac{\partial u'}{\partial y'} + \frac{\partial v'}{\partial x'}$$

Compare ε'_{xx} with the normal strain (along $\overline{CB}$) obtained earlier.

2.5 A cantilever beam of a rectangular cross-section is subjected to a shear force V as shown in Fig. 2.17. The bending stress is given by

$$\sigma_{xx} = \frac{Mz}{I}$$

where $M = -V(L-x)$. Assume a state of plane stress parallel to the x–z plane, i.e., $\sigma_{yy} = \tau_{xy} = \tau_{yz} = 0$. Find the transverse shear stress $\tau_{xz}(= \tau_{zx})$ by integrating the equilibrium equations over the beam thickness and applying the boundary conditions $\tau_{xz} = 0$ at $z = \pm h/2$. *Hint:* From the equilibrium equation

$$\frac{\partial \sigma_{xx}}{\partial x} + \frac{\partial \tau_{xz}}{\partial z} = 0$$

we have

$$\frac{\partial \tau_{xz}}{\partial z} = -\frac{\partial \sigma_{xx}}{\partial x} = -\frac{z}{I}\frac{\partial M}{\partial x}$$

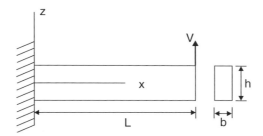

Figure 2.17 Cantilever beam subjected to a shear force.

2.6 The state of stress in a body is uniform and is given by

$$\sigma_{xx} = 4 \text{ MPa}, \qquad \tau_{xy} = 2 \text{ MPa}, \qquad \tau_{xz} = 0$$

$$\sigma_{yy} = 3 \text{ MPa}, \qquad \tau_{yz} = 0, \qquad \sigma_{zz} = 0$$

Find the three components of the stress vector **t** on the surface $ABCD$ as shown in Fig. 2.18. Find the normal component σ_n of the stress vector.

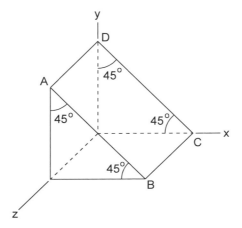

Figure 2.18   Shape of a wedge.

2.7 Find the principal stresses and corresponding principal directions for the stresses given in Problem 2.6. Check the result with other methods such as Mohr's circle.

2.8 A state of hydrostatic stress is given by

$$[\sigma] = \begin{bmatrix} \sigma_0 & 0 & 0 \\ 0 & \sigma_0 & 0 \\ 0 & 0 & \sigma_0 \end{bmatrix}$$

Show that on any surface the force (or stress vector) is always perpendicular to the surface and that the magnitude of the stress vector is equal to σ_0.

2.9 An isotropic solid with Young's modulus E and Poisson's ratio v is under a state of hydrostatic stress as given in Problem 2.8. Find the corresponding strain components.

2.10 For small strains, the volume change $\Delta V / V$ is identified to be equal to $\varepsilon_{xx} + \varepsilon_{yy} + \varepsilon_{zz}$. The bulk modulus K of an isotropic solid is defined as the ratio of the average stress and the volume change, i.e.,

$$\frac{1}{3}(\sigma_{xx} + \sigma_{yy} + \sigma_{zz}) = K \frac{\Delta V}{V}$$

Derive K in terms of E and v.

2.11 A block of elastic solid is compressed by normal stress σ_{xx} as shown in Fig. 2.19. The containing walls are rigid and smooth (frictionless). Find the values of k for plane strain and plane stress conditions, respectively, in the stress-strain relation obtained from the compression test above.

$$\sigma_{xx} = k\varepsilon_{xx}$$

Assume that $E = 70$ GPa and $\nu = 0.3$.

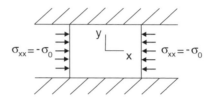

Figure 2.19   Solid between two smooth rigid walls.

2.12 An aluminum 2024 T3 bar of unit cross-sectional area is subjected to a tensile force in the longitudinal direction. If the lateral surface of the bar is confined and not allowed to contract during loading, find the force that is needed to produce a 1 percent longitudinal strain. Compare this with the corresponding load for the bar under simple tension.

2.13 Compare the axial stiffnesses of aluminum alloy 2024-T3 under plane strain and plane stress conditions, respectively.

2.14 Show that the state of stress of a solid body of any shape placed in a pressured chamber is a state of hydrostatic stress. Neglect the effect of the gravitational force.

2.15 Write the strain energy density expression in terms of stress components by using (2.95) for isotropic solids and show that the Poisson's ratio is bounded by -1 and 0.5.

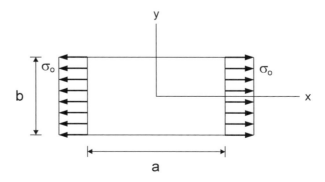

Figure 2.20   Thin rectangular panel subjected to uniform tension.

2.16 Derive the compatibility equation for plane elasticity problems in terms of stresses, i.e.,

$$\nabla^2(\sigma_{xx} + \sigma_{yy}) = 0$$

2.17 Consider a thin rectangular panel loaded as shown in Fig. 2.20. Show that the Airy stress function

$$\phi = c_1 x^2 + c_2 xy + c_3 y^2$$

solves the problem. Find the constants c_1, c_2, c_3.

2.18 Consider the $a \times b$ rectangular panel shown in Fig. 2.20. Find the problem that the Airy's stress function $\phi = xy^3$ solves. That is, find the tractions at the boundary of the panel.

3

TORSION

3.1 SAINT-VENANT'S PRINCIPLE

It is a common practice to adopt the resultant force and resultant moment rather than the actual traction in structural analyses. An example of a cantilever beam subjected to a shear force V is shown in Fig. 1.5. The actual application of the force could be quite different; namely, it could be the resultant of a distributed shear stress over the cross-section at the loading end or the sum of two concentrated forces applied at any two locations on the vertical centerline of the cross-section. We do not have any doubt about the consistency between the solutions obtained utilizing this simplifying approach. Indeed, such an idealization of loading conditions is justified by **Saint-Venant's principle**. According to Saint-Venant's principle, the stresses or strains at a point sufficiently far from the locations of two sets of applied loads do not differ significantly if these loads have the same resultant force and moment. These two sets of loads are said to be **statically equivalent**. An example of two statically equivalent tractions in a two-dimensional problem is given in Fig. 3.1. It is obvious that these two traction distributions have the same resultant force and resultant moment. Saint-Venant's principle asserts that the stresses or strains at a distant point in the body produced by these two loads, respectively, do not differ significantly. In general, the distance at which Saint-Venant's principle works is considered several times the size of the region of load application (the length b in this example). In fact, it also depends on the configuration of the body of interest.

Many solutions to structural mechanics problems are obtained by neglecting the boundary layer region in which the stress and strain fields are affected

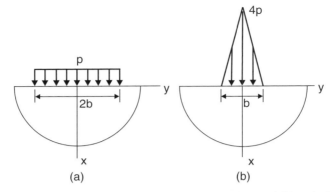

Figure 3.1 Elastic half-space subjected to (a) uniform traction, and (b) statically equivalent triangular traction.

by the actual local load distribution. It is of interest to know the size of this boundary layer in which these solutions are not accurate. In general, solutions for the boundary layer region are not easy to obtain analytically. Here, for illustration purposes, we use an idealized structure as an example.

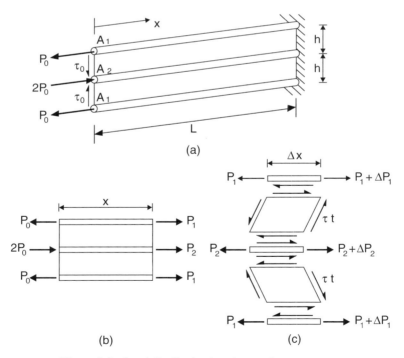

Figure 3.2 Load distribution in a three-stringer structure.

Consider a three-stringer thin-walled panel symmetrically loaded by a system of self-balanced forces as shown in Fig. 3.2a. The middle stringer has a cross-sectional area of A_2 and the upper and lower stringers have A_1 each. The webs are assumed to be capable of taking only shear stresses. That is, normal stresses in the webs are neglected.

From the free body diagram of Fig. 3.2b, we conclude that the forces in the stringers at section position x must satisfy

$$P_2 = -2P_1 \tag{3.1}$$

in which a negative sign means that the direction of P_2 is opposite to the assumed direction in Fig. 3.2b. Consider the free body diagram of a small segment Δx of the upper stringer at position x shown in Fig. 3.2c. The balance of forces in the axial direction yields

$$(P_1 + \Delta P_1) - P_1 - \tau t \, \Delta x = 0$$

in which t is the thickness of the web. By taking $\Delta x \to 0$, the equation above leads to

$$\tau = \frac{1}{t} \frac{dP_1}{dx} \tag{3.2}$$

Now take a strip of the upper web of length Δx at position x as shown in Fig. 3.3a before deformation. After the application of loads at the left end, shear strain occurs in the web and axial strains $\varepsilon_1(x)$ and $\varepsilon_2(x)$ occur in the upper and middle stringers, respectively (see Fig. 3.3b). The increment of shear strain is

$$\Delta \gamma = \frac{1}{h}(\varepsilon_1 \, \Delta x - \varepsilon_2 \, \Delta x)$$

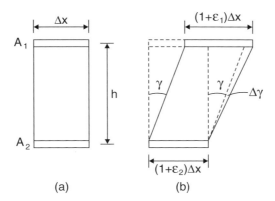

(a)   (b)

Figure 3.3   Free body diagrams of a strip of the upper web before and after deformation.

from which we obtain

$$\frac{d\gamma}{dx} = \frac{1}{h}(\varepsilon_1 - \varepsilon_2) \tag{3.3}$$

Since stringers are treated as axial members and webs are treated as shear panels, we have

$$\varepsilon_1 = \frac{P_1}{A_1 E}, \qquad \varepsilon_2 = \frac{P_2}{A_2 E}, \qquad \gamma = \frac{\tau}{G}$$

Substituting the relations above in (3.3), we obtain

$$\frac{d\tau}{dx} = \frac{G}{Eh}\left(\frac{P_1}{A_1} - \frac{P_2}{A_2}\right) \tag{3.4}$$

Substitution of (3.2) and (3.1) in (3.4) yields

$$\frac{d^2 P_1}{dx^2} - \lambda^2 P_1 = 0 \tag{3.5}$$

where

$$\lambda^2 = \frac{Gt}{Eh}\left(\frac{1}{A_1} + \frac{2}{A_2}\right) \tag{3.6}$$

The general solution for the second-order differential equation (3.5) is

$$P_1 = Ce^{-\lambda x} + De^{\lambda x} \tag{3.7}$$

Since the resultant of the applied forces is zero, it is expected that no force would be felt at a sufficiently large distance from the left end. Let L be large enough so that $P_1 \to 0$ as $x \to L$. This condition requires that $D = 0$. The other boundary condition is $P_1 = P_0$ at $x = 0$, which leads to $C = P_0$. To make (3.7) the solution, the shear stress τ_0 applied at the free end must be consistent with that given by (3.2), i.e.,

$$\tau_0 = \frac{1}{t}\frac{dP_1}{dx} = \frac{-\lambda P_0}{t}$$

The negative sign indicates that the actual direction of the shear stress must be opposite to that shown in Fig. 3.2.

For the case $A_1 = A_2 = A$, the solution of (3.7) can be written as

$$\frac{P_1}{P_0} = e^{-\lambda x} \tag{3.8}$$

where

$$\lambda = \sqrt{\frac{3Gt}{EAh}} = \frac{1}{h}\sqrt{\frac{3Ght}{EA}} \tag{3.9}$$

The distance x_0 from the free end at which $P_1/P_0 = 0.01$ can be obtained from solving (3.8) for x. We have

$$x = x_0 = \frac{6.9}{\lambda} \tag{3.10}$$

If the stringers and webs are made of aluminum 2024 with $E = 70$ GPa, $G = 27$ Gpa, $t = 2$ mm, $h = 200$ mm, $A_1 = A_2 = A = 100$ mm^2, then $\lambda = 0.0108$ mm^{-1}. Thus, from (3.10) we obtain $x_0 = 639$ mm. This says that Saint-Venant's principle is approximately valid beyond a distance that is three times the height h of the three-stringer panel from the load application end. Of course, this conclusion is affected by geometry of the structure. In general, thin-walled structures have small wall thicknesses and tend to make the value of λ small, resulting in a significant boundary layer in which Saint-Venant's principle is not valid.

3.2   TORSION OF UNIFORM BARS

Torque is a common form of load in aircraft structures. A torque is a moment or couple that has the unit N · m. The difference between a torque and a bending moment is that a torque acts about the longitudinal axis of a shaft as illustrated in Fig. 3.4, whereas a bending moment acts about an axis

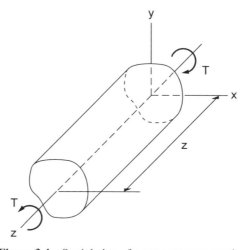

Figure 3.4   Straight bar of a constant cross-section.

that is perpendicular to the longitudinal axis of the shaft (beam). Based on Saint-Venant's principle, it is customary to ignore the differences in stress and strain near the load-application end of the shaft, even for statically equivalent loads. Thus, a torque symbol (a curved arrow) may represent many possible local load distributions that are statically equivalent. However, it should be noted that for shafts of thin-walled cross-sections under torsion, this boundary layer, in which stresses and strains are different for statically equivalent loads, may become large and, thus, needs to be taken into account in the stress analysis.[1]

The subject of torsion in a prismatic shaft of isotropic and linearly elastic solids is usually treated in the first course of mechanics of solid. However, only shafts of circular solid and hollow sections are considered. In deriving the deformation and stress fields in the shaft, the following assumptions are made:

- Plane sections of the shaft remain plane and circular after deformation produced by application of the torque.
- Diameters in plane sections remain straight after deformation

These assumptions lead to the result that shear strain (and, thus, shear stress) is a linear function of the radial distance from the point of interest to the center of the section. Moreover, these assumptions imply that plane sections of the shaft rotate as rigid bodies during deformation without in-plane deformations or out-of-plane displacements (i.e., warping). Unfortunately, these assumptions are not valid in shafts of noncircular sections, for which different formulations are required.

There are two classical approaches to solving the torsion of solid shafts of noncircular cross-section. The Prandtl stress function method employs assumptions regarding stresses produced in the shaft by a torque, while the Saint-Venant warping function method is based on assumptions of the displacement field. These two methods lead to the same solution. Here we begin with Saint-Venant's displacement assumptions but derive the governing equations in terms of the Prandtl stress function.

Consider a straight shaft of constant cross-section subjected to equal and opposite torques T at the ends as shown in Fig. 3.4. The origin of the coordinate system is selected to be at the **center of twist** (COT) of the cross-section, about which the cross-section rotates during twisting under the torque. By the definition of the COT, the in-plane displacements vanish at this location. For a circular cross-section, the COT is obviously located at the center of the cross-section. In general, the location of the COT depends on the shape of the cross-section and how the end is supported. In the formulation of the torsion problem, however, the explicit location of the COT may not be known a priori. If the effect of end support is ignored (i.e., warping is allowed to develop freely without exterior constraints), then the COT coincides with the shear center, which is discussed in Chapter 5. For convenience, we assume that the Cartesian coordinate system is set up with the origin located at the COT. The governing equations for torsion problems derived based on an arbitrarily selected origin of the coordinate system are identical to those based on this special coordinate system except for the displacements, which differ by a rigid body displacement. This result can be recognized by the fact that an arbitrary coordinate system is related to the COT system by a constant vector, which does not lead to nontrivial strains (and stresses).

Let α denote the total angle of rotation (twist angle) at z relative to the end at $z = 0$. The rate of twist (twist angle per unit length) at z is denoted by

$$\theta = \frac{\alpha}{z}$$

Saint-Venant assumed that during torsional deformation, plane sections warp, but their projections on the x–y plane rotate as a rigid body. This assumption implies that the in-plane displacement components u and v follow those of a rigid body rotation. Consider an arbitrary point P on the cross-section at z that moves through a rotation of a small angle α to P' after the torque is applied. For clarity in illustration, we select P to locate on the lateral boundary of the bar as shown in Fig. 3.5. Assume that the

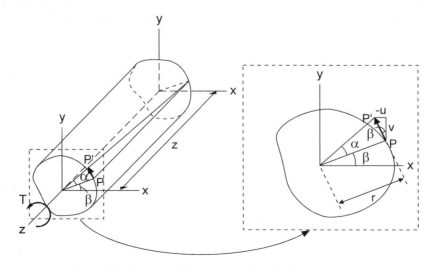

Figure 3.5   Rotation of the cross-section of a bar under torsion.

cross-section at $z = 0$ remains stationary. If the rotation angle α is small, then the displacement components at point P are given by

$$u = -r\alpha \sin \beta = -\alpha y = -\theta zy \qquad (3.11)$$

$$v = r\alpha \cos \beta = \alpha x = \theta zx \qquad (3.12)$$

in which r is the distance from the origin of the coordinates to point P. This displacement field represents a rigid rotation of the cross-section through angle α in the x–y plane.

The displacement w in the z-direction is assumed to be independent of z, and, thus, can be expressed in the form

$$w(x, y) = \theta \psi(x, y) \qquad (3.13)$$

where the rate of twist θ is independent of z and $\psi(x, y)$ is the warping function.

The displacement field given by (3.11)–(3.13) yields

$$\varepsilon_{xx} = \varepsilon_{yy} = \varepsilon_{zz} = \gamma_{xy} = 0$$

From the stress-strain relations we conclude that

$$\sigma_{xx} = \sigma_{yy} = \sigma_{zz} = \tau_{xy} = 0$$

Thus, τ_{yz} and τ_{xz} are the only two nonvanishing stress components. In view of the displacement field, it is easy to see that τ_{yz} and τ_{xz} are independent of z. In the absence of body forces, the equations of equilibrium (2.21)–(2.23) reduce to

$$\frac{\partial \tau_{xz}}{\partial x} + \frac{\partial \tau_{yz}}{\partial y} = 0 \qquad (3.14)$$

Prandtl introduced a stress function $\phi(x, y)$ such that

$$\tau_{xz} = \frac{\partial \phi}{\partial y}, \qquad \tau_{yz} = -\frac{\partial \phi}{\partial x} \qquad (3.15)$$

It is easy to verify that τ_{zx} and τ_{yz}, derived from ϕ in this manner, satisfy the equations of equilibrium automatically.

From (3.11) and (3.12) and the strain–displacement relations,

$$\gamma_{xz} = \frac{\partial w}{\partial x} + \frac{\partial u}{\partial z}, \qquad \gamma_{yz} = \frac{\partial w}{\partial y} + \frac{\partial v}{\partial z}$$

we obtain

$$\gamma_{xz} = \frac{\partial w}{\partial x} - \theta y \tag{3.16a}$$

$$\gamma_{yz} = \frac{\partial w}{\partial y} + \theta x \tag{3.16b}$$

Using (3.16), it is easy to derive the following equation:

$$\frac{\partial \gamma_{yz}}{\partial x} - \frac{\partial \gamma_{xz}}{\partial y} = 2\theta \tag{3.17}$$

This is the compatibility equation for torsion. Using the stress-strain relations

$$\gamma_{yz} = \frac{1}{G}\tau_{yz}, \qquad \gamma_{xz} = \frac{1}{G}\tau_{xz}$$

we obtain, from (3.17),

$$\frac{\partial \tau_{yz}}{\partial x} - \frac{\partial \tau_{xz}}{\partial y} = 2G\theta \tag{3.18}$$

In terms of the Prandtl stress function, (3.18) becomes

$$\frac{\partial^2 \phi}{\partial x^2} + \frac{\partial^2 \phi}{\partial y^2} = -2G\theta \tag{3.19}$$

The torsion problem now reduces to finding the stress function ϕ and requiring that the stresses derived from this stress function satisfy the boundary conditions.

On the lateral surface of the bar, no loads are applied. Thus, the stress vector (traction) **t** must vanish. Using (2.30), i.e.,

$$\{t\} = [\sigma]\{n\}$$

the stress vector can be evaluated on the lateral surface by specifying the unit normal vector **n**. On the lateral surface, $n_z = 0$. Thus,

$$\begin{Bmatrix} t_x \\ t_y \\ t_z \end{Bmatrix} = \begin{bmatrix} 0 & 0 & \tau_{xz} \\ 0 & 0 & \tau_{yz} \\ \tau_{xz} & \tau_{yz} & 0 \end{bmatrix} \begin{Bmatrix} n_x \\ n_y \\ 0 \end{Bmatrix} \tag{3.20}$$

Explicitly, we have

$$t_x = 0, \qquad t_y = 0 \tag{3.21a}$$

$$t_z = \tau_{xz}n_x + \tau_{yz}n_y = \frac{\partial \phi}{\partial y}n_x - \frac{\partial \phi}{\partial x}n_y \tag{3.21b}$$

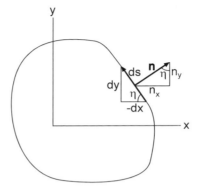

Figure 3.6 Tangential (s) and normal (n) directions of the boundary contour of the bar cross-section.

Referring to Fig. 3.6, it is easy to derive

$$n_x = \sin \eta = \frac{dy}{ds} \tag{3.22a}$$

$$n_y = \cos \eta = -\frac{dx}{ds} \tag{3.22b}$$

Using the relations in (3.22), (3.21b) can be expressed as

$$t_z = \frac{\partial \phi}{\partial y} \frac{dy}{ds} + \frac{\partial \phi}{\partial x} \frac{dx}{ds}$$

$$= \frac{d\phi}{ds} \tag{3.23}$$

The traction free boundary condition $t_z = 0$ is now given by

$$\frac{d\phi}{ds} = 0 \quad \text{or} \quad \phi = \text{constant} \tag{3.24a}$$

on the lateral surface. For solid sections with a single contour boundary, this constant is arbitrary and can be chosen to be zero. Thus, the boundary condition can be expressed as

$$\phi = 0 \quad \text{on the lateral surface of the bar} \tag{3.24b}$$

Of interest are the shear stresses τ_{xz} and τ_{yz} on the cross-section (see Fig. 3.7) and their resultant torque. Consider a differential area $dA = dx\,dy$. The torque produced by the stresses in this area is

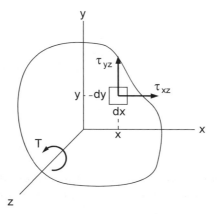

Figure 3.7   Shear stresses on the cross-section.

$$dT = x\tau_{yz}\, dA - y\tau_{xz}\, dA$$

$$= \left(-x\frac{\partial\phi}{\partial x} - y\frac{\partial\phi}{\partial y}\right) dA$$

The total resultant torque is obtained by integrating dT over the entire cross-section, i.e.,

$$T = -\iint_A \left(x\frac{\partial\phi}{\partial x} + y\frac{\partial\phi}{\partial y}\right) dx\, dy$$

$$= -\iint_A \left[\frac{\partial}{\partial x}(x\phi) - \phi\right] dx\, dy - \iint_A \left[\frac{\partial}{\partial y}(y\phi) - \phi\right] dx\, dy$$

$$= 2\iint_A \phi\, dx\, dy - \int [x\phi]_{x_1}^{x_2}\, dy - \int [y\phi]_{y_1}^{y_2}\, dx$$

where x_1, x_2, y_1, and y_2 are integration limits on the boundary. Since ϕ vanishes on the boundary contour, the last two terms in the equation above vanish. Thus,

$$T = 2\iint_A \phi\, dx\, dy \tag{3.25}$$

The derivations above clearly indicate that the solution of the torsion problem lies in finding the stress function $\phi(x, y)$ that vanishes along the lateral boundary of the bar. Once $\phi(x, y)$ is determined, the location of the center of twist $(x = 0, \ y = 0)$ is also defined.

For bars of arbitrary cross-section, warping (out-of-plane displacement) of the cross-section occurs when twisted. From (3.16) we have

$$\frac{\partial w}{\partial x} = \frac{\tau_{xz}}{G} + \theta y \qquad \text{and} \qquad \frac{\partial w}{\partial y} = \frac{\tau_{yz}}{G} - \theta x$$

The warping displacement w can be obtained by integrating the equations above.

The torsion equation (1.10) can be used to express the relation between the applied torque T and the resulting rate of twist θ for shafts of arbitrary cross-sections. The torsion constant J is obtatined as

$$J = \frac{T}{G\theta}$$

Using (3.25) and (3.19) in the equation above, we obtain

$$J = -\frac{4}{\nabla^2 \phi} \iint\limits_{A} \phi \, dx \, dy$$

Thus, once the Prandtl stress function is solved, the **torsional rigidity** GJ of the shaft is also determined.

3.3   BARS WITH CIRCULAR CROSS-SECTIONS

Consider a uniform bar of circular cross-section. If the origin of the coordinates is chosen to coincide with the center of the cross-section, the boundary contour is given by the equation

$$x^2 + y^2 = a^2$$

where a is the radius of the circular boundary. Assume the stress function as

$$\phi = C \left(\frac{x^2}{a^2} + \frac{y^2}{a^2} - 1 \right) \tag{3.26}$$

This stress function satisfies the boundary condition (3.24).

Substituting (3.26) into the compatibility equation (3.19), we have

$$C = -\tfrac{1}{2}a^2 G\theta \tag{3.27}$$

Thus, the stress function of (3.26) with C given by (3.27) solves the torsion problem. It also indicates that the center of the circular section is the center of twist.

From (3.25), we have the torque as

$$T = 2C \iint\limits_A \left(\frac{x^2}{a^2} + \frac{y^2}{a^2} - 1 \right) dx \, dy$$

$$= 2C \iint\limits_A \left(\frac{r^2}{a^2} - 1 \right) dA$$

$$= 2C \left(\frac{I_p}{a^2} - A \right)$$

where

$$I_p = \iint\limits_A r^2 \, dA = \frac{1}{2} \pi a^4$$

is the polar moment of inertia of the cross-section, and

$$A = \pi a^2$$

is the cross-sectional area. Since $a^2 A = 2I_p$, thus,

$$T = -\frac{2C I_p}{a^2} = \theta G I_p \tag{3.28}$$

and $J = I_p$. The shear stresses are

$$\tau_{xz} = \frac{\partial \phi}{\partial y} = 2C \frac{y}{a^2} = -G\theta y \tag{3.29}$$

$$\tau_{yz} = -\frac{\partial \phi}{\partial x} = -2C \frac{x}{a^2} = G\theta x \tag{3.30}$$

Consider a cylinder of a circular cross-section of radius r cut from the circular bar of radius a. On the lateral surface of this cylinder of radius r (see Fig. 3.8a), the stress vector is given by (3.21). Thus,

$$t_x = t_y = 0$$

$$t_z = \tau_{xz} n_x + \tau_{yz} n_y$$

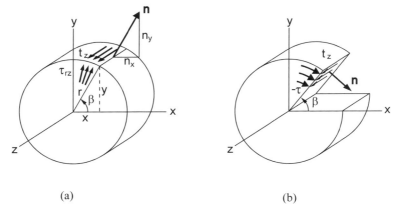

Figure 3.8   Shear stresses on (a) a cylinder of radius $r\,(a > r)$ cut out of the original cylinder, and (b) a surface cut along the radial direction of the cylinder.

Also note that

$$n_x = \cos \beta = \frac{x}{r}$$

$$n_y = \sin \beta = \frac{y}{r}$$

Using (3.29), (3.30), and the relations above, we obtain

$$t_z = -G\theta \frac{xy}{r} + G\theta \frac{xy}{r} = 0$$

As shown in Fig. 3.8a, the radial shear stress τ_{rz} on the cross-section vanishes since it is equal to t_z.

Now, consider the surface exposed by cutting along the radial direction of the cylinder as shown in Fig. 3.8b. The unit normal vector to the surface is given by

$$n_z = 0, \quad n_x = \sin \beta = \frac{y}{r}, \quad n_y = -\cos \beta = -\frac{x}{r} \qquad (3.31)$$

Substituting (3.29) and (3.30) together with (3.31) into (3.21b) yields the only nonvanishing component of the stress vector in the z-direction as

$$t_z = -G\theta r$$

On the z-face (the cross-section), the tangential shear stress τ (that is perpendicular to the radial direction) is equal to t_z in magnitude. Adjusting the sign for direction, we have

$$\tau = -t_z = G\theta r$$

Using (3.28) to eliminate θ, the relation above can be expressed in terms of the torque as

$$\tau = \frac{Tr}{J}$$

It is evident that the magnitude of τ is proportional to r. This is the well-known result for torsion of circular bars.

Using (3.16), (3.29), (3.30), and stress-strain relations we can also show that

$$w = 0$$

Thus, for bars with circular cross-sections under torsion, there is no warping.

3.4 BARS WITH NARROW RECTANGULAR CROSS-SECTIONS

The result for circular cross-sections cannot be extended automatically to noncircular cross-sections. For example, for a square cross-section, the shear stress cannot be assumed to be perpendicular to the radial direction, and its magnitude is not proportional to the radial distance. Further, warping is present. For bars with certain noncircular cross-sections, solutions can be found in books on the theory of elasticity. In aircraft structures, many components are large in lateral dimensions compared with the thickness. For such narrow sections, simplifications can be achieved.

Consider a bar (shaft) with a narrow rectangular cross-section subjected to a pure torque as shown in Fig. 3.9a. From the consideration of symmetry, the center of twist for this cross-section is located at the geometric center of the section. To satisfy Saint-Venant's principle, the length L of the bar is assumed to be much greater than the width b of the cross-section. Moreover, it is assumed that the cross-section (see Fig. 3.9b) of the bar is wide such that $b \gg t$. Basically, this is a thin plate long in the z-direction. On the top and bottom faces ($y = \pm t/2$), the traction-free boundary condition requires that

$$\tau_{yz} = 0$$

In terms of the stress function, this says that

$$\frac{\partial \phi}{\partial x} = -\tau_{yz} = 0 \tag{3.32}$$

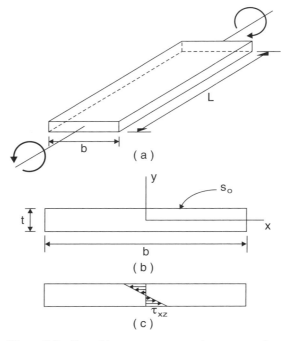

Figure 3.9 Bar with a narrow rectangular cross-section.

on the top and bottom faces. Since t is very small, and τ_{yz} must vanish at $y = \pm t/2$, it is unlikely that the shear stress τ_{yz} would build up across the thickness. Therefore, we can assume that $\tau_{yz} \approx 0$ through the thickness. Consequently, we assume that ϕ is independent of x.

In view of the foregoing, the governing equation (3.19) reduces to

$$\frac{d^2\phi}{dy^2} = -2G\theta \tag{3.33}$$

Integrating (3.33) twice, we obtain

$$\phi = -G\theta y^2 + C_1 y + C_2$$

The boundary condition requires that

$$\phi = 0 \quad \text{at} \quad y = \pm\frac{t}{2} \tag{3.34}$$

which leads to

$$C_1 = 0, \qquad C_2 = G\theta\frac{t^2}{4}$$

and subsequently,

$$\phi = -G\theta \left(y^2 - \frac{t^2}{4} \right) \tag{3.35}$$

The corresponding shear stresses are obtained from (3.15) and (3.35) as

$$\tau_{xz} = -2G\theta y, \qquad \tau_{yz} = 0 \tag{3.36}$$

The shear stress τ_{xz} acts parallel to the x-axis and is distributed linearly across the width, as shown in Fig. 3.9c. The maximum shear stress occurs at $y = \pm t/2$, i.e.,

$$(\tau_{xz})_{\text{max}} = G\theta t \tag{3.37}$$

The torque is obtained by substituting (3.35) into (3.25):

$$T = -2G\theta \int_{-b/2}^{b/2} \int_{-t/2}^{t/2} \left(y^2 - \frac{t^2}{4} \right) dy\, dx$$

$$= \frac{bt^3}{3} G\theta$$

Define the torsion constant J as

$$J = \frac{bt^3}{3} \tag{3.38}$$

Then

$$T = GJ\theta$$

where GJ is the torsional rigidity.

From (3.16a) and (3.36), we have

$$\frac{\partial w}{\partial x} = \gamma_{xz} + \theta y = \frac{\tau_{xz}}{G} + \theta y = -\theta y$$

The amount of warping on the cross-section can be obtained from integrating the expression above. We obtain

$$w = -xy\theta$$

The integration constant is set equal to zero because $w = 0$ at the center of twist. In fact, $w = 0$ along the centerline of the sheet.

The results obtained here can be used for sections composed of a number of thin-walled members. For example, the T-section shown in Fig. 3.10a

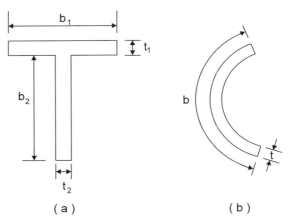

Figure 3.10   Examples of open thin-walled sections.

can be considered as a section consisting of two rectangular sections. The combined torsional rigidity is given by

$$GJ = G(J_1 + J_2)$$

where

$$J_1 = \tfrac{1}{3}b_1 t_1^3$$
$$J_2 = \tfrac{1}{3}b_2 t_2^3$$

The formula given by (3.38) can also be applied to curved open thin-walled sections by interpreting b as the total arc length as depicted by Fig. 3.10b.

It is noted that the torsion constant J given by (3.38) is valid only if b/t is large. If the two dimensions are comparable, then J should be evaluated using the elasticity solution obtained by solving a rectangular cross-section using the Pradtl stress function method.[2] A correction coefficient β needs to modify the torsion constant of (3.38), i.e.,

$$J = \beta \frac{bt^3}{3}$$

For $b/t = 1.0, 1.5, 2.0, 5.0, 10.0$, $\beta = 0.422, 0.588, 0.687, 0.873, 0.936$, respectively.

---

[2]S. P. Timoshenko and J. N. Goodier, *Theory of Elasticity*, 3rd ed., McGraw-Hill, New York, 1970, p. 309.

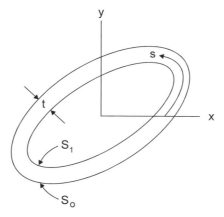

Figure 3.11   Wall section enclosed by an inner contour S_1 and an outer contour S_0.

3.5 CLOSED SINGLE-CELL THIN-WALLED SECTIONS

Members with closed thin-walled sections are quite common in aircraft structures. Figure 3.11 shows a closed thin-walled section with a single cell. The wall thickness t is assumed to be small compared with the total length of the complete wall contour. In general, the wall thickness t is not a constant but is a function of s.

The wall section is enclosed by the inner contour S_1 and the outer contour S_0 as shown in Fig. 3.11. Using the Prandtl stress function ϕ, the stress-free boundary conditions are given by [see (3.24)]

$$\frac{d\phi}{ds} = 0 \quad \text{on } S_1 \text{ and } S_0 \tag{3.39}$$

Thus,

$$\phi = C_0 \quad \text{on } S_0 \tag{3.40}$$

$$\phi = C_1 \quad \text{on } S_1 \tag{3.41}$$

where C_0 and C_1 are two different constants and cannot be set equal to zero simultaneously as in the case of solid sections with a single boundary contour.

Consider the shear stresses at an arbitrary point on the wall section. Let us set up a coordinate system s–n so that s coincides with the centerline of the wall and n is perpendicular to s as shown in Fig. 3.12a. Take an infinitesimal prismatic element of unit length in the z-direction as shown in Fig. 3.12b. The active shear stresses on the side faces are shown in the figure. Note that the inclined surface is perpendicular to the s-direction. The equilibrium

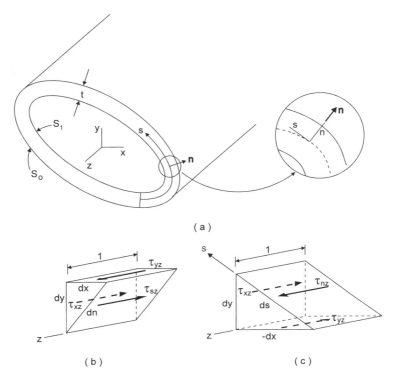

(a)

(b)   (c)

Figure 3.12 Shear stresses at a point on a wall section.

condition (balance of forces in the z-direction) gives

$$\tau_{sz} \, dn = -\tau_{xz} \, dy + \tau_{yz} \, dx$$

$$\tau_{sz} = -\tau_{xz} \frac{\partial y}{\partial n} + \tau_{yz} \frac{\partial x}{\partial n}$$

$$= -\frac{\partial \phi}{\partial y} \frac{\partial y}{\partial n} - \frac{\partial \phi}{\partial x} \frac{\partial x}{\partial n}$$

$$= -\frac{\partial \phi}{\partial n} \tag{3.42}$$

Similarly, using the free body of Fig. 3.12c and the equilibrium condition, we have

$$\tau_{nz} \, ds = \tau_{xz} \, dy - \tau_{yz} \, dx$$

$$\tau_{nz} = \tau_{xz} \frac{\partial y}{\partial s} - \tau_{yz} \frac{\partial x}{\partial s}$$

$$= \frac{\partial \phi}{\partial y} \frac{\partial y}{\partial s} + \frac{\partial \phi}{\partial x} \frac{\partial x}{\partial s}$$

$$= \frac{\partial \phi}{\partial s} \tag{3.43}$$

Note that a negative sign is added in front of $\tau_{yz}\, dx$ to account for the fact that an increment ds is accompanied by a decrement $-dx$. Since $\tau_{nz} = \partial \phi / \partial s = 0$ on S_0 and S_1, and t is small, the variation of τ_{nz} across the wall thickness is negligible. Hence, a reasonable approximation is to assume $\tau_{nz} \approx 0$ over the entire wall section. As a result of this assumption, the τ_{sz} is retained as the only nonvanishing stress component.

Let ϕ be expressed in terms of the coordinates s and n and expand ϕ in series of n as

$$\phi(s, n) = \phi_0(s) + n\phi_1(s) + n^2\phi_2(s) + \cdots \tag{3.44}$$

where

$$-\frac{t}{2} \leq n \leq \frac{t}{2}$$

in which $t(s)$ is wall thickness and, in general, is a function of the wall contour.

Since the range of n is small, the high-order terms of n in (3.44) can be neglected without causing much error. Retaining the linear term in (3.44), we have

$$\phi(s, n) = \phi_0(s) + n\phi_1(s) \tag{3.45}$$

The boundary conditions require that

$$\phi\left(s, \frac{t}{2}\right) = \phi_0 + \frac{t}{2}\phi_1 = C_0 \quad \text{on } S_0$$

$$\phi\left(s, -\frac{t}{2}\right) = \phi_0 - \frac{t}{2}\phi_1 = C_1 \quad \text{on } S_1$$

Solving the two equations, we obtain

$$\phi_0 = \frac{1}{2}(C_0 + C_1)$$

$$\phi_1 = \frac{1}{t}(C_0 - C_1)$$

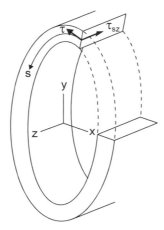

Figure 3.13   Shear stress on a wall section in the s-direction.

The shear stress τ on the wall section in the s-direction (see Fig. 3.13) is given by

$$\tau = \tau_{sz} = -\frac{\partial \phi}{\partial n} = -\phi_1 = \frac{1}{t}(C_1 - C_0) \tag{3.46}$$

Thus, the shear stress τ on the thin-walled section is uniform over the thickness. Nevertheless, τ is still a function of the contour s if the wall thickness t is not constant.

Define the **shear flow** q (force/contour length) as

$$q = \tau t = C_1 - C_0 \tag{3.47}$$

This indicates that regardless of the wall thickness, the shear flow is constant along the wall section.

The shear stress τ along the wall is usually represented by the shear flow q along the centerline of the wall. Since the shear flow forms a closed contour, the force resultants are equal to zero, i.e., $\sum F_x = 0$ and $\sum F_y = 0$. However, the shear flow produces a resultant torque.

Consider a constant shear flow q on a closed thin-walled section as shown in Fig. 3.14. The resultant torque produced by the shear flow acting on the contour segment ds is given by

$$dT = \rho q \, ds$$

where ρ is the distance from the origin of the coordinates to the line segment ds. The torque T can be obtained by integrating the above along the entire shear flow contour. Noting that $\rho \, ds = 2 \, dA$, we rewrite the integral above

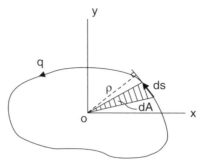

Figure 3.14   Constant shear flow on a closed thin-walled section.

as an area integral:

$$T = \oint \rho q \, ds = \iint_{\overline{A}} 2q \, dA = 2q\overline{A} \tag{3.48}$$

where $\overline{A}$ is the area enclosed by the shear flow or, equivalently, *the area enclosed by the centerline of the wall section.*

Consider a shear flow q as shown in Fig. 3.15. It can easily be shown that the resultant force **R** is oriented parallel to the line connecting the two end points P and Q of the shear flow, and the magnitude of the resultant force is given by

$$R = qd \tag{3.49a}$$

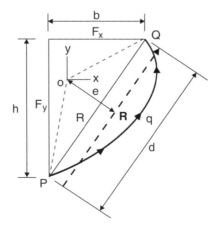

Figure 3.15   Resultants and moment of a constant shear flow.

The components of the resultant force are

$$F_x = qb \tag{3.49b}$$

$$F_y = qh \tag{3.49c}$$

The torque about the z-axis is

$$T = 2\overline{A}q \tag{3.49d}$$

where $\overline{A}$ is the area bounded by the contour q and lines $\overline{OP}$ and $\overline{OQ}$. The actual location e (see Fig. 3.15) of the resultant force can be obtained from the torque equivalence condition, i.e.,

$$Re = T = 2\overline{A}q \tag{3.50}$$

Twist Angle: Method 1 Using the shear strains given by (3.16) and the stress-strain relations, we obtain

$$\tau_{xz} = G\left(\frac{\partial w}{\partial x} - \theta y\right) \tag{3.51a}$$

$$\tau_{yz} = G\left(\frac{\partial w}{\partial y} + \theta x\right) \tag{3.51b}$$

Using the first line of (3.42), we have

$$\tau = \tau_{sz} = -\tau_{xz}\frac{\partial y}{\partial n} + \tau_{yz}\frac{\partial x}{\partial n} \tag{3.52}$$

From Fig. 3.16, the following relations are obvious:

$$\frac{dy}{dn} = -\frac{dx}{ds}, \qquad \frac{dx}{dn} = \frac{dy}{ds} \tag{3.53}$$

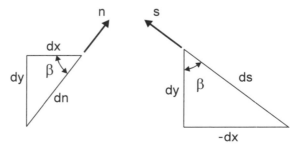

Figure 3.16 Geometrical relations among coordinate increments.

Substitution of (3.51) and (3.53) into (3.52) leads to

$$
\begin{aligned}
\tau &= G \left(\frac{\partial w}{\partial x} - \theta y \right) \frac{\partial x}{\partial s} + G \left(\frac{\partial w}{\partial y} + \theta x \right) \frac{\partial y}{\partial s} \\
&= G \left(\frac{\partial w}{\partial x} \frac{\partial x}{\partial s} + \frac{\partial w}{\partial y} \frac{\partial y}{\partial s} \right) + G\theta \left(x \frac{\partial y}{\partial s} - y \frac{\partial x}{\partial s} \right) \\
&= G \frac{\partial w}{\partial s} + G\theta \left(x \frac{\partial y}{\partial s} - y \frac{\partial x}{\partial s} \right)
\end{aligned} \tag{3.54}
$$

Integrating τ over the closed contour along the centerline of the wall, we have

$$
\begin{aligned}
\oint \tau \, ds &= G \oint \frac{\partial w}{\partial s} \, ds + G\theta \oint \left(x \frac{\partial y}{\partial s} - y \frac{\partial x}{\partial s} \right) ds \\
&= G w |_0^L + G\theta \oint (x \, dy - y \, dx)
\end{aligned} \tag{3.55}
$$

where L is the total length of the contour. The first term on the right-hand side of (3.55) vanishes because $w(0) = w(L)$. The second term can be integrated using Green's theorem, which states that

$$
\oint (f \, dx + g \, dy) = \iint_{\overline{A}} \left(\frac{\partial g}{\partial x} - \frac{\partial f}{\partial y} \right) dA
$$

By identifying g as x and f as $-y$ in (3.55), we use Green's theorem to obtain

$$
\oint \tau \, ds = 2G\theta \overline{A}
$$

from which θ is obtained as

$$
\theta = \frac{1}{2G\overline{A}} \oint \tau \, ds = \frac{1}{2G\overline{A}} \oint \frac{q}{t} \, ds \tag{3.56}
$$

Twist Angle: Method 2 for Constant Shear Flow Consider a thin-walled bar of unit length as shown in Fig. 3.17. The shear stress is

$$
\tau = \frac{q}{t}
$$

and the shear strain is

$$
\gamma = \frac{\tau}{G} = \frac{q}{Gt}
$$

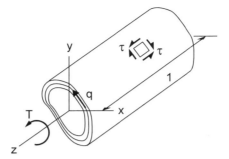

Figure 3.17   Thin-walled bar of unit length.

The corresponding strain energy density is given by

$$W = \frac{1}{2}\tau\gamma = \frac{q^2}{2Gt^2}$$

The total strain energy stored in the bar (of unit length) is

$$U = \oint Wt\,ds$$

$$= \oint \frac{q^2}{2Gt}\,ds \tag{3.57}$$

The work done by torque T through the twist angle θ is given by

$$W_e = \tfrac{1}{2}T\theta = \tfrac{1}{2}\cdot 2q\overline{A}\theta = q\overline{A}\theta$$

In the equation above, $\overline{A}$ is the area enclosed by the shear flow contour.

From the energy principle (work done by external forces is equal to the total strain energy), we obtain

$$W_e = U$$

or explicitly,

$$q\overline{A}\theta = \frac{q^2}{2G}\oint \frac{ds}{t}$$

Thus,

$$\theta = \frac{q}{2\overline{A}G}\oint \frac{ds}{t} \tag{3.58}$$

This is identical to (3.56) if q is constant along the wall.

Since $T = GJ\theta$, we have

$$\frac{T}{GJ} = \frac{q}{2\overline{A}G} \oint \frac{ds}{t}$$

From this relation, we obtain the torsion constant J for the single-cell thin-walled section as

$$J = \frac{2\overline{A}T}{q \oint ds/t}$$

$$= \frac{4\overline{A}^2}{\oint ds/t} \tag{3.59}$$

in which $T = 2q\overline{A}$ has been used.

Example 3.1 Consider a thin-walled tube with the cross-section shown in Fig. 3.18a. The wall thickness is $t = 0.005$ m and the average radius is 0.2025 m. Thus,

$$\overline{A} = \pi(0.2025)^2 = 0.129 \text{ m}^2$$

$$\oint \frac{ds}{t} = \frac{\pi \times 0.405}{0.005} = 254$$

From (3.59), the torsion constant is obtained as

$$J_1 = \frac{4\overline{A}^2}{\oint ds/t} = \frac{4 \times (0.129)^2}{254} = 2.62 \times 10^{-4} \text{ m}^4$$

If the tube is cut open as shown in Fig. 3.18b, then the torsion constant is given by (3.38) as

$$J_2 = \frac{bt^3}{3} = \frac{\pi \times 0.4 \times (0.005)^3}{3} = 5.24 \times 10^{-8} \text{ m}^4$$

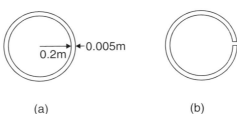

(a)   (b)

Figure 3.18   Thin-walled tube with (a) a closed section, and (b) a slit section.

The ratio of torsional rigidities of these two tubes is

$$\frac{J_1}{J_2} = 5000$$

It is evident that the tube with the closed section has a much higher torsional rigidity than the slit tube.

Example 3.2   Consider a three-stringer thin-walled beam with the cross-section as shown in Fig. 3.19. The contribution of individual stringers to the overall torsional rigidity of the thin-walled structure is small and can be neglected. Hence, this structure can be considered as a single-cell closed section with a nonuniform wall thickness and the shear flow is constant along the wall.

If the torque T (N·m) is given, then the shear flow is obtained from the relation $T = 2\overline{A}q$. The area $\overline{A}$ is readily obtained as

$$\overline{A} = \tfrac{1}{2}\pi(0.6)^2 + \tfrac{1}{2}(2 \cdot 1.2) = 1.765 \text{ m}^2 \tag{a}$$

Thus,

$$q = \frac{T}{2\overline{A}} = \frac{T}{3.53} \qquad \text{N/m}$$

The twist angle per unit length can be obtained using (3.56). We have

$$\theta = \frac{1}{2\overline{A}G}q \oint \frac{ds}{t}$$

$$= \frac{q}{2 \times 1.765G}\left(\frac{1.2\pi}{2t_1} + \frac{2}{t_2} + \frac{2.33}{t_3}\right)$$

$$= 79\frac{T}{G} \qquad \text{rad/m} \tag{b}$$

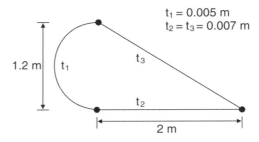

$t_1 = 0.005$ m
$t_2 = t_3 = 0.007$ m

1.2 m t_1 t_3 t_2 2 m

Figure 3.19   Three-stringer thin-walled bar.

Because of its smaller thickness, the shear stress in the curved wall is higher than that in the straight walls. The shear stress in the curved wall is

$$\tau = \frac{q}{t_1} = \frac{T}{0.005 \times 3.53} = 56.66T$$

If the allowable shear stress of the material is 200 MPa, then the maximum torque that this structure can take is

$$T_{max} = \frac{\tau_{allow}}{56.66} = \frac{200 \times 10^6}{56.66} = 3.53 \times 10^6 \text{ N·m}$$

3.6 MULTICELL THIN-WALLED SECTIONS

Wing sections are often composed of airfoil skin supported by thin vertical webs to form multicell constructions. Figure 3.20 shows a two-cell skin–web section. In addition, stiffeners are used to carry bending loads. *The individual stiffeners, although having large concentrated cross-sectional areas, have relatively small torsion constants and do not make a significant contribution to the torsional rigidity of the wing box and are often neglected in the consideration of torsional stiffness of the wing box.*

For torsion of a single-cell section, the Prandtl stress function must be constant along each boundary contour. For the two-cell section, there are three boundary contours, i.e., S_0, S_1, and S_2 (see Fig. 3.21). Thus, we have

$$\phi(S_0) = C_0$$
$$\phi(S_1) = C_1$$
$$\phi(S_2) = C_2$$

where C_0, C_1, and C_2 are three different constants.

From the result of (3.47), we note that the shear flow between two boundary contours is equal to the difference between the values of ϕ along these

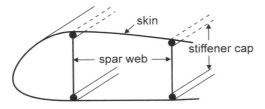

Figure 3.20   Two-cell stringer–skin–web section.

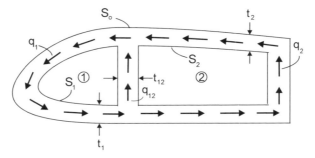

Figure 3.21 Two-cell thin-walled section.

contours. Specifically, for each cell, the shear flow is considered positive if it forms a counterclockwise torque about the cell and its value is equal to the value of ϕ on the inside contour minus that on the outside contour, i.e.,

$$q_1 = C_1 - C_0 \tag{3.60a}$$

$$q_2 = C_2 - C_0 \tag{3.60b}$$

$$q_{12} = C_1 - C_2 \tag{3.60c}$$

Note that the direction of q_{12} assumed in Fig. 3.21 is positive for cell 1 and negative for cell 2.

From (3.60), we obtain

$$q_{12} = q_1 - q_2 \tag{3.61}$$

In view of this relation, the shear flow system of Fig. 3.21 can be viewed as the superposition of two shear flows q_1 and q_2 as depicted in Fig. 3.22. Thus, the torque contributed by each cell can be calculated by using $T = 2\overline{A}q$. The total torque of the two-cell section is

$$T = 2\overline{A}_1 q_1 + 2\overline{A}_2 q_2 \tag{3.62}$$

where $\overline{A}_1$ and $\overline{A}_2$ are the areas enclosed by the shear flows q_1 and q_2, respectively.

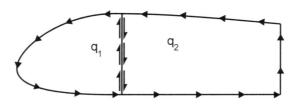

Figure 3.22 Superposition of two constant shear flows.

The twist angles θ_1 and θ_2 of the cells are obtained using (3.56):

$$\theta_1 = \frac{1}{2\overline{A}_1 G} \oint_{\text{cell 1}} \frac{q\, ds}{t} \tag{3.63a}$$

$$\theta_2 = \frac{1}{2\overline{A}_2 G} \oint_{\text{cell 2}} \frac{q\, ds}{t} \tag{3.63b}$$

It is important to note that for each cell, the shear flow (and twist angle) is taken positive if it flows (rotates) in the counterclockwise direction. For the shear flow in Fig. 3.21, $q_{12} = q_1 - q_2$ should be used for cell 1, while for cell 2, $-q_{12} = q_2 - q_1$ should be used.

Since the entire thin-wall section must rotate as a rigid body in the plane, we require the compatibility condition

$$\theta_1 = \theta_2 = \theta \tag{3.64}$$

Equations (3.62) and (3.64) are solved to find the two unknown shear flows q_1 and q_2.

Sections with more than two cells can be treated in a similar way. Additional equations provided by the compatibility condition are available for solving additional unknown shear flows. The torque for an n-cell section is given by

$$T = \sum_{i=1}^{n} 2\overline{A}_i q_i$$

The twist angle θ of the section is the same as the individual cells. Hence, we choose cell 1 to calculate the twist angle:

$$\theta = \theta_1 = \frac{1}{2\overline{A}_1 G} \oint_{\text{cell 1}} \frac{q\, ds}{t}$$

From $T = GJ\theta$, we obtain the torsion constant J of the multicell section as

$$J = \frac{T}{G\theta} = \frac{4\overline{A}_1 \sum \overline{A}_i q_i}{\oint_{\text{cell 1}} q\, ds/t} \tag{3.65}$$

Example 3.3   A two-cell thin-walled box beam (see Fig. 3.23) is subjected to a torque T that causes a twist angle $\theta = 5°/\text{m}$ (0.087 rad/m). Assume that $G = 27$ GPa.

Using (3.56), we have for cell 1,

$$\theta_1 = \frac{1}{2G\overline{A}_1} \left[\frac{q_1(0.4 + 0.5 + 0.4)}{t_1} + \frac{q_{12}(0.5)}{t_{12}} \right]$$

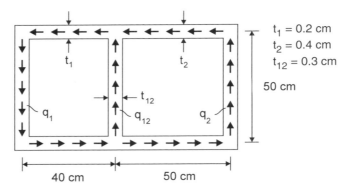

Figure 3.23 Two-cell thin-walled box beam.

Substituting numerical values into the equation above, we have

$$0.087 = 7.56 \times 10^{-8} q_1 - 1.55 \times 10^{-8} q_2 \tag{a}$$

Similarly for cell 2,

$$\theta_2 = \frac{1}{2G\bar{A}_2} \left[\frac{q_2(1.5)}{t_2} - \frac{(q_1 - q_2)(0.5)}{t_{12}} \right]$$

$$0.087 = -1.24 \times 10^{-8} q_1 + 4.01 \times 10^{-8} q_2 \tag{b}$$

Solving (a) and (b), we obtain the shear flows as

$$q_1 = 1.7 \times 10^6 \text{ N/m}$$

$$q_2 = 2.7 \times 10^6 \text{ N/m}$$

The torque that produces the given twist angle is

$$T = 2\bar{A}_1 q_1 + 2\bar{A}_2 q_2 = 2.03 \times 10^6 \text{ N·m}$$

The torsion constant is

$$J = \frac{T}{G\theta} = \frac{2.03 \times 10^6}{(27 \times 10^9) \times 0.087} = 0.86 \times 10^{-3} \text{ m}^4$$

Example 3.4 The cross-section of the thin-walled beam in Fig. 3.19 is modified into the two-cell section shown in Fig. 3.24. It is of interest to compare the torsional rigidities of these two beams.

For the single-cell beam of Fig. 3.19, the torsional rigidity can be obtained from Eq. (b) in Example 3.2, from which we have

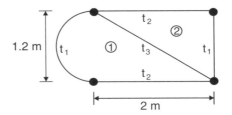

Figure 3.24 Dimensions of a two-cell section.

$$T = \frac{1}{79}G\theta = 0.013G\theta \ \text{N·m} \tag{a}$$

and the torsion constant $J = 0.013 \ \text{m}^4$.

For the two-cell section of Fig. 3.24, we have $\overline{A}_1 = 1.767 \ \text{m}^2$ and $\overline{A}_2 = 1.2 \ \text{m}^2$ for cells 1 and 2, respectively. From (3.62), we obtain

$$T = 3.53q_1 + 2.4q_2 \tag{b}$$

in which the shear flows q_1 and q_2 are indicated in Fig. 3.25. The rate of twist for cell 1 is

$$\theta_1 = \frac{1}{2 \times 1.765G} \left[\frac{1.2\pi}{2t_1}q_1 + \frac{2}{t_2}q_1 + \frac{2.33}{t_3}(q_1 - q_2) \right] \tag{c}$$

Similarly, for cell 2 we have

$$\theta_2 = \frac{1}{2 \times 1.2G} \left[\frac{2.33}{t_3}(q_2 - q_1) + \frac{1.2}{t_1}q_2 + \frac{2}{t_2}q_2 \right]$$

Finally, the compatibility equation (3.64) leads to

$$\frac{1}{1.765} \left[\frac{1.2\pi}{2t_1}q_1 + \frac{2}{t_2}q_1 + \frac{2.33}{t_3}(q_1 - q_2) \right]$$
$$= \frac{1}{1.2} \left[\frac{2.33}{t_3}(q_2 - q_1) + \frac{1.2}{t_1}q_2 + \frac{2}{t_2}q_2 \right] \tag{d}$$

Figure 3.25 Shear flows in a two-cell section.

Solving (b) and (d) simultaneously for q_1 and q_2, we obtain

$$q_1 = 0.17T, \qquad q_2 = 0.16T$$

Since the two cells have identical twist angle, we can use (c) to calculate the rate of twist, with the result

$$\theta = \theta_1 = \frac{32.9}{G}T \qquad \text{or} \qquad T = 0.03G\theta$$

Thus, the two-cell section has a torsion constant $J = 0.03 \text{ m}^4$, which is more than twice that of the single-cell section of Fig. 3.19.

Consider the single-cell section obtained from that of Fig. 3.24 by removing the diagonal sheet. The shear flow is easily obtained from

$$T = 5.93q$$

Subsequently, the rate of twist is obtained as

$$\theta = \frac{33.8}{G}T$$

and the torsion constant is

$$J = 0.0296 \text{ m}^4$$

This value is almost equal to that of the two-cell section. From this example, it is interesting to note that tortional rigidity of a closed thin-walled section cannot be increased significantly by compartmentalizations.

3.7 WARPING IN OPEN THIN-WALLED SECTIONS

Except for circular cross-sections, shafts of noncircular sections exhibit warping under pure torques. In other words, out-of-plane displacements occur during torsion. For instance, in the narrow rectangular section shown in Fig. 3.9, the out-of-plane displacement is given by

$$w = -xy\theta \tag{3.66}$$

Substituting (3.66) in (3.16b), we have

$$\gamma_{yz} = 0 \tag{3.67}$$

which is consistent with initial assumption $\tau_{yz} = 0$. Moreover, we note that $\tau_{xz} = 0$ along the centerline of the wall from (3.36). Thus, we have

$$\gamma_{xz} = 0 \tag{3.68}$$

along $y = 0$.

It is noted that $w = 0$ along the centerline ($y = 0$) of the wall. Thus, warping occurs only across the thickness of the wall. This type of warping is usually called **secondary warping**. For general thin-walled sections, the centerline of the wall may also warp with the magnitude much greater than the secondary warping. This is known as **primary warping**.

Consider a curved thin-walled section of uniform thickness as shown in Fig. 3.26. Following the procedure described in Section 3.5, we set up a right-hand coordinate system $s-n-z$ so that s coincides with the centerline of the wall, n perpendicular to s, and z remains unchanged. The origin of s can be chosen arbitrarily. At any local position along the contour s, the state of stress (or strain) in the wall section is approximately the same as that of a straight rectangular section by identifying s with x and n with y, respectively (see Fig. 3.9b). However, the warping function cannot be obtained from (3.66) by using such a substitution. At a point along the s contour, let u_s denote the displacement in the s-direction and u_n the displacement in the n-direction. With respect to this new coordinate system, the shear strains are

$$\gamma_{sz} = \frac{\partial w}{\partial s} + \frac{\partial u_s}{\partial z}, \qquad \gamma_{nz} = \frac{\partial w}{\partial n} + \frac{\partial u_n}{\partial z} \tag{3.69}$$

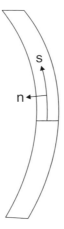

Figure 3.26 $s-n$ coordinates along the centerline of the thin-walled section.

In view of (3.68), we have $\gamma_{sz} = 0$ along s (the centerline of the wall). Thus,

$$\frac{\partial w}{\partial s} = -\frac{\partial u_s}{\partial z} \tag{3.70}$$

Apparently, the out-of-plane displacement along s may be obtained by integration of (3.70) when $\partial u_s / \partial z$ is known.

For convenience, we assume that the center of twist for the thin-walled section is known. Figure 3.27 shows the relative position of the center of twist and the contour s. Take an arbitrary point P on the contour s. After the application of a torque, point P moves to P' because of the rigid rotation assumption of Saint-Venant's torsion formulation (see Fig. 3.5). The total displacement is

$$\overline{PP'} = r\alpha = rz\theta$$

where α is the total twist angle measured from $z = 0$ to the current section of interest.

From Fig. 3.27a, the displacement u_s in the s-direction (i.e., the tangential direction at point P) is

$$u_s = \overline{PP'} \cos \beta = rz\theta \cos \beta = \rho z\theta \tag{3.71}$$

where ρ is the distance from the center of twist to the tangent line at point P as shown in Fig. 3.27a. Substituting (3.71) in (3.70), we obtain

$$\frac{\partial w}{\partial s} = -\rho\theta \tag{3.72}$$

To perform the integration of (3.72) along s, we consider a line segment ds along the contour s as shown in Fig. 3.27b. The area $d\overline{A}$ is recognized

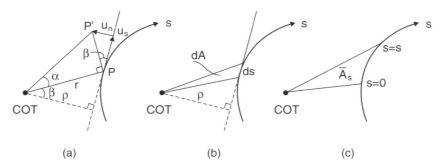

Figure 3.27 Calculation of warping along a thin-walled section.

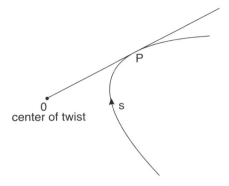

Figure 3.28   Positive direction of an s contour.

to be

$$d\overline{A} = \tfrac{1}{2}\rho\,ds \tag{3.73}$$

Now, integrating (3.72) with the aid of (3.73) we obtain

$$w(s) - w(0) = -\theta \int_0^s \rho\,ds = -\theta \iint_{A_s} 2\,d\overline{A} = -2\overline{A}_s\theta \tag{3.74}$$

where $\overline{A}_s$ is the area enclosed by the contour s and the two lines connecting the center of twist with the two points $s = 0$ and $s = s$, respectively (see Fig. 3.27c). The area $\overline{A}_s$ can also be considered as the area swept by the generator line (the line connecting COT and the origin of s) from $s = 0$ to $s = s$. If the origin of s is selected such that warping vanishes at $s = 0$, then $w(0) = 0$.

It is important to note that in the derivation of (3.71), the positive direction of s must be set up so that it is counterclockwise with respect to the z-axis in order to make it consistent with the positive direction of θ. Otherwise, a negative sign must be added on the right-hand side of (3.71). This can be implemented by interpreting the area $\overline{A}_s$ to be a negative value. As an example, consider the s contour shown in Fig. 3.28. Line $\overline{OP}$ is tangential to the s curve at point P. Below point P, the positive direction of s forms a counterclockwise rotation about the z-axis, while beyond point P it forms a clockwise rotation. Thus, in the calculation of $\overline{A}_s$, the additional area beyond point P must be regarded as a negative value. On the other hand, if s is set up such that its positive direction is opposite to that shown in Fig. 3.28, then the area $\overline{A}_s$ is positive above point P and negative below point P.

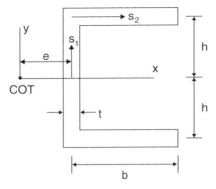

Figure 3.29 Thin-walled channel section.

Example 3.5 The center of twist of the thin-walled channel section shown in Fig. 3.29 is located on the horizontal axis of symmetry at a distance

$$e = \frac{tb^2h^2}{I_x}$$

to the left of the vertical wall, where I_x is the moment of inertia about the x-axis. Break up the contour s into two straight parts, s_1 and s_2, as shown in Fig. 3.29. For contour s_1 we have

$$\overline{A}_{s_1} = \tfrac{1}{2}es_1$$

and for contour s_2

$$\overline{A}_{s_2} = -\tfrac{1}{2}hs_2$$

in which a negative sign is added because the direction of s_2 produces a clockwise rotation about the center of twist. From (3.74), the warp of the thin wall is

$$w(s_1) = w(0) - 2A_{s_1}\theta = w(0) - es_1\theta \qquad \text{on } s_1$$

and

$$w(s_2) = w(s_1 = h) - 2\overline{A}_{s_2}\theta = w(0) - eh\theta + hs_2\theta \qquad \text{on } s_2$$

The warping displacement in the lower half of the section can easily be recognized to be of the same magnitude as the upper half except for the sign change resulting from the fact that the s contour is in the negative direction. In view of the antisymmetric warping of the section, we conclude that $w(0) = 0$. In fact, this is true at locations on any axis of symmetry of the thin-walled section. Thus, the warping displacement at the upper corner point is $-eh\theta$ (a negative sign means that the displacement is in the negative z-direction), and at the upper free edge it is $(bh - eh)\theta$.

3.8 WARPING IN CLOSED THIN-WALLED SECTIONS

We follow the same procedure used in treating open sections. First, set up the s–n–z coordinate system with the origin at the COT and with positive s forming counterclockwise about the COT (see Fig. 3.30). The shear flow q_s is related to the shear stress τ_{sz} as

$$q_s = t\tau_{sz} = tG\gamma_{sz} \tag{3.75}$$

where t is the thickness of the wall and G is the shear modulus. Using the first equation in (3.69), we obtain from (3.75) the following relation:

$$q_s = Gt\left(\frac{\partial w}{\partial s} + \frac{\partial u_s}{\partial z}\right) \tag{3.76}$$

Using (3.71), which is valid for closed sections, we obtain from (3.76) the equation

$$\frac{\partial w}{\partial s} = \frac{q_s}{Gt} - \rho\theta \tag{3.77}$$

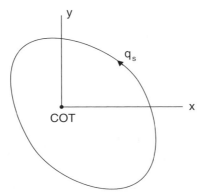

Figure 3.30   Shear flow on a closed thin-walled section.

Integration of (3.77) along s leads to

$$w(s) - w(0) = \int_0^s \frac{q_s}{Gt} \, ds - \theta \int_0^s \rho \, ds$$

$$= \int_0^s \frac{q_s}{Gt} \, ds - 2\overline{A}_s \theta \tag{3.78}$$

This gives the warp at any point relative to that at the point $s = 0$.

Example 3.6 The cross-section of a thin-walled box beam subjected to a torque T is shown in Fig. 3.31. Find the warp of the cross-section.

The axes of symmetry of this section are chosen to coincide with x and y axes, respectively, as shown in Fig. 3.31. From symmetry, the COT is seen to be at the origin of the coordinate system. Moreover, we take the origin for s at $(x = a/2, y = 0)$. From symmetry, we have $w(0) = 0$. In fact, any midpoint of the four sidewalls can be selected as the origin of s and satisfies $w(0) = 0$.

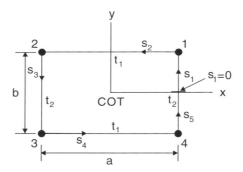

Figure 3.31 Thin-walled box beam with four stringers.

For closed single-celled sections, we have

$$q_s = \frac{T}{2\overline{A}} \tag{a}$$

and

$$\theta = \frac{1}{2G\overline{A}} \oint \frac{q_s}{t} \, ds = \frac{T}{4G\overline{A}^2} \oint \frac{ds}{t} \tag{b}$$

in which $\overline{A}$ is the total area enclosed by the thin walls. Substituting (a) and (b) in (3.78), we have the warp along the thin wall as

$$w(s) = \frac{T}{2G\overline{A}} \left(\int_0^s \frac{ds}{t} - \frac{\overline{A}_s}{\overline{A}} \oint \frac{ds}{t} \right)$$

$$= \frac{T}{2Gab} \left[\int_0^s \frac{ds}{t} - 2\overline{A}_s \left(\frac{1}{bt_1} + \frac{1}{at_2} \right) \right] \tag{c}$$

By dividing the contour s into five consecutive segments, s_1, s_2, s_3, s_4, and s_5, the integration in (c) can easily be evaluated in each segment. For s_1, we obtain

$$w(s_1) = \frac{T}{2Gab} \left[\frac{s_1}{t_2} - \frac{as_1}{2} \left(\frac{1}{bt_1} + \frac{1}{at_2} \right) \right]$$

which is linear in s_1. At stringer 1, $s_1 = b/2$ and the warp is

$$w_1 = w\left(\frac{b}{2}\right) = \frac{T}{8Gab}\left(\frac{b}{t_2} - \frac{a}{t_1}\right) \tag{d}$$

The warp along s_2 is

$$w(s_2) = w_1 + \frac{T}{2Gab}\left[\frac{s_2}{t_1} - \frac{bs_2}{2}\left(\frac{1}{bt_1} + \frac{1}{at_2}\right)\right]$$

The warp at stringer 2 is obtained by setting $s_2 = a$. We have

$$w_2 = -w_1$$

Similarly, the warps along s_3, s_4, and s_5 can be calculated using (c). The result would show that

$$w_2 = w_4 = -w_1 = -w_3$$

Of course, the result above can also be deduced from symmetry of the cross-section. Note that a positive value of w_1 means that the warp is in the positive z-direction if T is positive (counterclockwise). It is obvious from (d) that the sign of w_1 changes if $a/t_2 > b/t_1$ and that no warping occurs for square sections of uniform wall thickness.

3.9 EFFECT OF END CONSTRAINTS

The Saint-Venant torsion formulation is derived based on the assumption that warping is freely developed and is uniform along the shaft. However, in practice, torsion members may be connected to other structural components or built into a "rigid" support. As a result, this assumption is violated and the Saint-Venant torsion solutions obtained in previous sections need to be modified. Nowadays, analyses of the end constraint effects in structures under general loads are often carried out using finite elements. In this section we consider shafts of open thin-walled sections to illustrate the effect of end constraints in torsion. The purpose is to gain some insight into how suppression of warping can increase the torsional rigidity of a shaft. It is noted that the following analysis procedures are not suitable for thin-walled structures with stringers. If stringeers are present, their contributions to the shear flow induced by end constraints must be included.

Consider an open thin-walled section in which an s contour is set up along the thin wall as shown in Fig. 3.28. The solution for free warping is given by (3.74). If we select the origin of s such that $w(0) = 0$, then (3.74) becomes

$$w(s) = -2\overline{A}_s \theta \qquad (3.79)$$

In the case of torsion without end constraints, both w and θ are independent of z over the entire length of the shaft. If end constraints are present, then both w and θ are functions of z (the longitudinal axis along the shaft). Let us write (3.79) in the form

$$w(s, z) = -w_s(s)\,\theta(z) \qquad (3.80)$$

in which

$$w_s(s) = 2\overline{A}_s \qquad (3.81)$$

is the warp at any point s per unit rate of twist angle. From (3.80) we note that w_s is independent of z, meaning that the "shapes" of the warp on different cross-sections are identical except for the magnitude of the warp. The fact that $w(s, z)$ is a function of z indicates the presence of the longitudinal normal stress:

$$\sigma_{zz}(z, s) = E\varepsilon_{zz}(z, s) = E\frac{\partial w}{\partial z} = -Ew_s(s)\frac{d\theta}{dz} \qquad (3.82)$$

which in turn induces shear flow q as shown in Fig. 3.32. Since σ_{zz} is a function of z, its values at the two cross-sections separated by a distance dz differ by the increment $d\sigma_{zz}$. Similarly, the shear flows on the two longitudinal sides of the free body are q and $q + dq$. The balance of forces in the z-direction acting on the free body yields

$$(\sigma_{zz} + d\sigma_{zz})t\,ds + (q + dq)\,dz - \sigma_{zz}t\,ds - q\,dz = 0$$

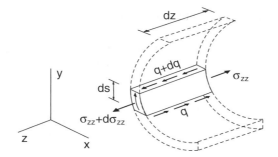

Figure 3.32 Shear flow induced by longitudinal normal stress.

or

$$dq = -\frac{d\sigma_{zz}}{dz} t \, ds \qquad (3.83)$$

Substituting (3.82) in (3.83), we have

$$dq = E w_s \frac{d^2\theta}{dz^2} t \, ds \qquad (3.84)$$

The shear flow along a certain contour on a wall can be obtained by integration (3.84) and using the fact that shear flow vanishes at the open edges, say, $s = s_0$ and $s = s_1$, of the section. Then at any location s, the shear flow is

$$q(s) = E \frac{d^2\theta}{dz^2} \int_{s_0}^{s} w_s t \, ds \qquad (3.85)$$

In the case where multiple walls are present, the shear flow along each path must be calculated in the same manner. The shear flow obtained from (3.85) is produced by σ_{zz}. This shear flow forms a torque T' about the center of twist (COT) which can be expressed as

$$T' = \int_{s_0}^{s_1} \rho q \, ds \qquad (3.86)$$

where the definition of ρ is given in Fig. 3.27a. Again, in the case of multiple thin-walled contours, the integration in (3.86) must include all integrals along all the s contours.

Substituting of (3.85) in (3.86) yields

$$T' = E \frac{d^2\theta}{dz^2} \int_{s_0}^{s_1} \rho \left(\int_{s_0}^{s} w_s t \, ds \right) ds \qquad (3.87)$$

From (3.73) and (3.81), we obtain the relation

$$\rho = 2\frac{d\overline{A}_s}{ds} = \frac{dw_s}{ds} \tag{3.88}$$

Thus,

$$T' = E\frac{d^2\theta}{dz^2}\int_{s_0}^{s_1}\frac{dw_s}{ds}\left(\int_{s_0}^{s} w_s t\, ds\right) ds \tag{3.89}$$

Noting that

$$\frac{d}{ds}\left(w_s\int_{s_0}^{s} w_s t\, ds\right) = \frac{dw_s}{ds}\left(\int_{s_0}^{s} w_s t\, ds\right) + w_s\frac{d}{ds}\int_{s_0}^{s} w_s t\, ds$$

we have

$$\frac{dw_s}{ds}\left(\int_{s_0}^{s} w_s t\, ds\right) = \frac{d}{ds}\left(w_s\int_{s_0}^{s} w_s t\, ds\right) - w_s^2 t\, ds \tag{3.90}$$

Substituting (3.90) in (3.89) and integrating, we obtain

$$T' = E\frac{d^2\theta}{dz^2}\left[w_s\int_{s_0}^{s} w_s t\, ds\right]_{s_0}^{s_1} - E\frac{d^2\theta}{dz^2}\int w_s^2 t\, ds \tag{3.91}$$

Since shear flows vanish at open edges of the shaft, i.e., $q(s_0) = q(s_1) = 0$, the first term on the right-hand side of (3.91) is zero in view of (3.85). Thus, the torque induced by the end constraint is

$$T' = -E\Gamma\frac{d^2\theta}{dz^2} \tag{3.92}$$

where

$$\Gamma = \int_{s_0}^{s_1} w_s^2 t\, ds = \int_{s_0}^{s_1} 4\overline{A}_s^2 t\, ds \tag{3.93}$$

is a constant that depends only on the geometry of the cross-section. Note that the area $\overline{A}_s$ must be measured starting from the point where warp is zero. Since $\overline{A}_s$ appears as a squared term in (3.93), the direction of the contour s is not relevant.

Based on the superposition principle, the total torque carried by the shaft is the sum of the torque of the Saint-Venant torsion and T'. Thus,

$$T = GJ\theta - E\Gamma\frac{d^2\theta}{dz^2} \tag{3.94}$$

which can be written in the form

$$\theta - k^2 \frac{d^2\theta}{dz^2} = \frac{T}{GJ} \tag{3.95}$$

In (3.95),

$$k^2 = \frac{E\Gamma}{GJ} \tag{3.96}$$

Note that k has the length unit. If the applied torque T is known, then the general solution to the second-order differential equation is

$$\theta = C_1 e^{z/k} + C_2 e^{-z/k} + \theta_p \tag{3.97}$$

where θ_p is a particular solution that depends on the functional form of T, and C_1 and C_2 are arbitrary constants to be determined by boundary conditions.

For a constant torque T, a particular solution is $\theta_p = T/GJ$, and the solution (3.97) can be written in the form

$$\theta = \frac{T}{GJ} \left(1 + D_1 e^{z/k} + D_2 e^{-z/k} \right) \tag{3.98}$$

Assume that the shaft is built in at $z = 0$ and at the free end $z = L$ a torque T is applied. Let L be large such that $T = GJ\theta$ (Saint-Venant torsion) as $z \to L$. To satisfy this condition, we require that $D_1 = 0$. At the built-in end ($z = 0$), warping is suppressed and $w = 0$. From (3.80), this implies that $\theta = 0$. Thus, $D_2 = -1$ and

$$\theta = \frac{T}{GJ} \left(1 - e^{-z/k} \right) = \frac{T}{GJ_{\text{eff}}}$$

where the effective torsion constant J_{eff} is given by

$$J_{\text{eff}} = \frac{J}{1 - e^{-z/k}}$$

It is noted that the effective torsional rigidity GJ_{eff} is a function of z and its value is enhanced significantly as z approaches the built-in end. Also note that the end constraint effect diminishes as $k/z \to 0$. Thus, k is a length parameter that measures the size of the boundary layer of the end constraint effect.

Once the solution for θ is obtained, the distribution of the end constraint–induced normal stress σ_{zz} can be calculated by using (3.82) and the shear flow by using (3.85).

Example 3.7   The two ends of an aluminimum I-beam are rigidly welded, respectively, to two heavy steel plates to suppress warping at both ends (see Fig. 3.33a). A torque T is applied by twisting one end against the other. From symmetry of the cross-section, it is easy to see that the center of twist is located at the midpoint of the web. At this location the warping displacement is zero. This location is then selected as the origin of the s contour.

The I-beam section can be approximated by three thin rectangular sections and the torsion constant J is given by

$$J = \frac{2bt^3 + ht^3}{3} = 4.27 \times 10^{-9} \text{ m}^4$$

To calculate Γ using (3.93), we must perform integration over the entire section. We note that $\overline{A}_s$ for the web is zero and that the four flange segments have the same value of $\overline{A}_s$ except the sign. In view of the foregoing, we consider only the upper right segment of the flange, and a local s' contour (Fig. 3.33b) is set up for the evaluation of $\overline{A}_s$. The swept area about the center of twist is

$$\overline{A}_s = -\frac{1}{2}\left(\frac{h}{2}\right)s' = -0.025s' \qquad 0 \le s' \le \frac{b}{2}$$

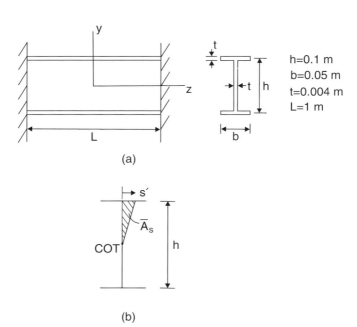

(a)

(b)

Figure 3.33   I-beam with two built-in ends.

The contribution of this segment to the Γ of the entire section is

$$\int_0^{b/2} 4\overline{A}_s^2 t \, ds = \int_0^{0.025} 4(0.025s')^2(0.004) \, ds = 52 \times 10^{-12} \text{ m}^6$$

The Γ for the entire section is the sum of the four equal contributions of the four flange segments. We have

$$\Gamma = 4 \times 52 \times 10^{-12} = 208 \times 10^{-12} \text{ m}^6$$

Taking $E = 70$ GPa and $G = 27$ GPa for aluminum, then from (3.96),

$$k = \sqrt{\frac{E\Gamma}{GJ}} = 0.355 \text{ m}$$

The general solution (3.98) can be written alternatively as

$$\theta = \frac{T}{GJ}\left(1 + B_1 \cosh\frac{z}{k} + B_2 \sinh\frac{z}{k}\right) \tag{a}$$

Because of symmetry of θ with respect to z, the odd function $\sinh(z/k)$ should be dropped. This is accomplished by setting $B_2 = 0$. The second boundary condition is at the built-in end ($z = L/2$), where warp $w = w_s\theta$ is zero, i.e., $\theta = 0$ or

$$0 = \frac{T}{GJ}\left(1 + B_1 \cosh\frac{L}{2k}\right)$$

from which we obtain

$$B_1 = -\frac{1}{\cosh(L/2k)}$$

Since $L = 1$ m and $k = 0.335$ m, we have $B_1 = -0.428$. Thus, the solution for rate of twist angle is

$$\theta = \frac{T}{GJ}\left(1 - 0.428\cosh\frac{z}{k}\right) \tag{b}$$

The twist angle α at any point z is

$$\alpha = \int_0^z \theta \, dz = \frac{T}{GJ}\int_0^z \left(1 - 0.428\cosh\frac{z}{k}\right) dz$$

$$= \frac{T}{GJ}\left(z - 0.143\sinh\frac{z}{k}\right) \tag{c}$$

At the built-in end ($z = L/2$), the twist angle relative to that at midspan ($z = 0$) is calculated by substituting $z = L/2$ in (c), with the result

$$\alpha = 0.198 \frac{T}{GJ}$$

If the end constraint is removed, the twist angle is calculated using the Saint-Venant torsion formulation, with the result

$$\alpha = \frac{T}{GJ}$$

The end constraint effect on the amount of twist in the shaft is evident. In other words, suppression of warping in a shaft increases its torsional rigidity.

PROBLEMS

3.1 Show that there is no warping in a bar of circular cross-section.

3.2 Show that the Prandtl stress function for bars of circular solid sections is also valid for bars of hollow circular sections as shown in Fig. 3.34. Find the torsion constant J in terms of the inner radius a_i and outer radius a_0, and compare with the torsion constant obtained using (3.59) for thin-walled sections. What is the condition on the wall thickness for the approximate J to be within 1 percent of the exact J?

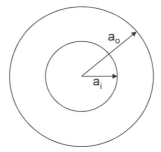

Figure 3.34   Bar of a hollow circular section.

3.3 Consider the straight bar of a uniform elliptical cross-section. The semimajor and semiminor axes are a and b, respectively. Show that the stress function of the form

$$\phi = C \left(\frac{x^2}{a^2} + \frac{y^2}{b^2} - 1 \right)$$

provides the solution for torsion of the bar.

Find the expression of C and show that

$$J = \frac{\pi a^3 b^3}{a^2 + b^2}$$

$$\tau_{zx} = \frac{-2Ty}{\pi ab^3}, \qquad \tau_{zy} = \frac{2Tx}{\pi a^3 b}$$

and the warping displacement

$$w = \frac{T(b^2 - a^2)}{\pi a^3 b^3 G} xy$$

3.4 A thin aluminum sheet is to be used to form a closed thin-walled section. If the total length of the wall contour is 100 cm, what is the shape that would achieve the highest torsional rigidity? Consider elliptical (including circular), rectangular, and equilateral triangular shapes.

3.5 The two-cell section in Fig. 3.35 is obtained from the single-cell section of Fig. 3.36 by adding a vertical web of the same thickness as the skin.

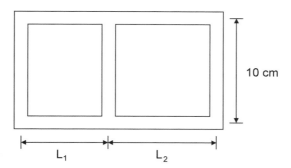

Figure 3.35   Two-cell thin-walled section.

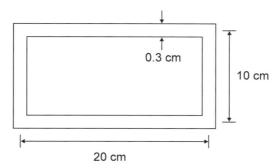

Figure 3.36   Single-cell section.

Compare the torsional rigidities of the structures of Figs. 3.35 and 3.36 with $L_1 = L_2 = 10$ cm and $L_1 = 5$ cm and $L_2 = 15$ cm, respectively.

3.6 Find the torsional rigidity if the sidewall of one of the two cells in Fig. 3.35 (with $L_1 = L_2 = 10$ cm) is cut open. What is the reduction of torsional rigidity compared with the original intact structure?

3.7 Find the torque capability of the thin-walled bar with the section shown in Fig. 3.36. Assume that the shear modulus $G = 27$ GPa and the allowable shear stress of $\tau_{\text{allow}} = 187$ MPa.

3.8 A two-cell thin-walled member with the cross-section shown in Fig. 3.37 is subjected to a torque T. The resulting twist angle θ is $3°$. Find the shear flows, the applied torque, and the torsion constant. The material is aluminum alloy 2024-T3.

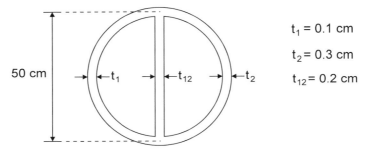

$t_1 = 0.1$ cm

$t_2 = 0.3$ cm

$t_{12} = 0.2$ cm

Figure 3.37   Two-cell section.

3.9 For the bar of Fig. 3.37, find the maximum torque if the allowable shear stress is $\tau_{\text{allow}} = 187$ MPa. What is the corresponding maximum twist angle θ?

3.10 Find the shear flow and twist angle in the two-cell three-stringer thin-walled bar with the cross-section shown in Fig. 3.38. The material is Al 2024-T3. The applied torque is 2×10^5 N·m.

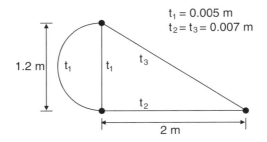

$t_1 = 0.005$ m

$t_2 = t_3 = 0.007$ m

Figure 3.38   Two-cell three-stringer thin-walled section.

3.11 What is the maximum torque for the structure of Fig. 3.38 if the allowable twist angle θ is $2°/m$?

3.12 The two shafts of thin-walled cross-sections shown in Fig. 3.39a and b, respectively, contain the same amount of aluminum alloy. Compare the torsional rigidities of the two shafts without end constraints.

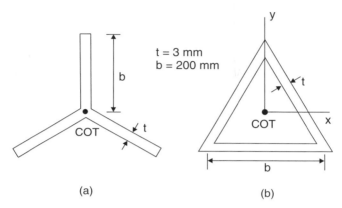

(a)   (b)

Figure 3.39 Cross-sections of two shafts.

3.13 Find the distributions of the primary warping displacement on the cross-sections shown in Fig. 3.39b. Due to symmetry, the center of twist coincides with the centroid of the section, and warp at the midpoint of each flat sheet section is zero. Sketch the warping displacement along the wall.

3.14 A shaft with a channel section shown in Fig. 3.40 is subjected to a torque T. Assume that neither end is constrained. Find the warping distribution on the cross-section, the maximum warp, and the location of the maximum warp.

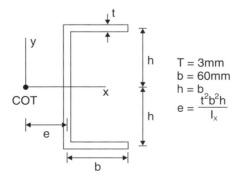

Figure 3.40 Dimensions of a channel section.

3.15 Consider the shaft of the channel section shown in Fig. 3.40. If one end of the shaft is built in and the other end is free, find the effective torsional rigidity as a function of the distance from the built-in end. Assume that the length L of the shaft is sufficiently large so that near the free end the Saint-Venant torsion assumptions are valid. Compare the total twist angle with that for a free–free shaft for $L = 2$ m.

3.16 Calculate the distributions of normal stress σ_{zz} and shear flow distributions at the built-in end for Problem 3.15.

3.17 Assume that the shaft of the channel section of Fig. 3.40 is built in at both ends. Find the torque that is necessary to produce a relative twist angle $\alpha = 5°$ between the two ends. Assume that $L = 1$ m, Young's modulus $E = 70$ GPa, and shear modulus $G = 27$ GPa. Compare this with the free–free case.

4

BENDING AND FLEXURAL SHEAR

4.1 DERIVATION OF THE SIMPLE (BERNOULLI–EULER) BEAM EQUATION

Consider a straight beam (bar) of a uniform cross-section that is symmetrical with respect to a vertical line. The coordinates are set up such that the x-axis coincides with the centroidal axis of the cross-sections along the beam, and the z-axis coincides with the vertical line of symmetry; see Fig. 4.1. The resultant transverse load p_z (N/m) is applied in the x–z plane.

Of interest are displacements in the x and z directions, u and w, respectively. If the width of the beam is small, then the state of stress due to transverse loading can be approximated by plane stress parallel to the x–z plane, and u and w can be assumed to be functions of x and z only. Expand u and w in power series of z as

$$u(x, z) = u_0(x) + zu_1(x) + z^2u_2(x) + \cdots \qquad (4.1)$$

$$w(x, z) = w_0(x) + zw_1(x) + z^2w_2(x) + \cdots \qquad (4.2)$$

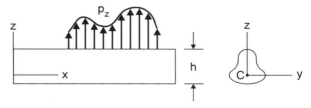

Figure 4.1 Straight beam of a uniform and symmetrical cross-section.

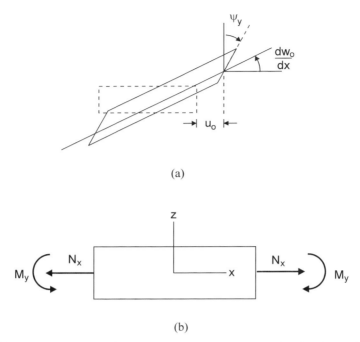

(a)

(b)

Figure 4.2 (a) Deformation of a beam element; (b) sign convention for the resultant force and moment.

For slender beams, the depth is small compared with the length. In other words, the range of z is small, and the high-order terms in z make insignificant contributions. Hence, as a first-order approximation, we truncate the series of (4.1) and (4.2) as

$$u = u_0(x) + z\psi_y(x) \tag{4.3}$$

$$w = w_0(x) \tag{4.4}$$

in which we use ψ_y in place of u_1. From (4.3), it is obvious that u_0 represents the longitudinal displacement at the centroidal axis, and ψ_y represents the rotation of the cross-section after deformation; see Fig. 4.2a. From (4.3), a positive rotation ψ_y is clockwise, which is opposite to the slope dw_0/dx of the beam deflection.

Note that $u(x, z)$ is a linear function of z. This implies that plane cross-sections remain plane after deformation but may not be perpendicular to the centroidal axis.

The strain components corresponding to the approximate displacements given by (4.3) and (4.4) are

$$\varepsilon_{xx} = \frac{\partial u}{\partial x} = \frac{du_0}{dx} + z\frac{d\psi_y}{dx} \tag{4.5}$$

$$\gamma_{xz} = \frac{\partial w}{\partial x} + \frac{\partial u}{\partial z} = \frac{dw_0}{dx} + \psi_y \tag{4.6}$$

Define the resultant axial force N_x and bending moment M_y as

$$N_x = \iint_A \sigma_{xx}\, dA \tag{4.7}$$

$$M_y = \iint_A z\sigma_{xx}\, dA \tag{4.8}$$

in which the area integration is over the entire cross-section. The definitions of N_x and M_y given by (4.7) and (4.8), respectively, also determine the sign convention for N_x and M_y. The positive directions of N_x and M_y are shown in Fig. 4.2b.

For slender beams, the transverse shear strain γ_{xz} is small. In calculating the bending strain, we can assume that $\gamma_{xz} = 0$ as an approximation. This leads to, from (4.6),

$$\psi_y = -\frac{dw_0}{dx} \tag{4.9}$$

The relation above implies that the plane cross-section remains perpendicular to the centroidal axis after deformation, and that the amount of rotation of the cross-section is equal to the slope of deflection.

Using (4.9) and (4.5), we obtain

$$\sigma_{xx} = E\varepsilon_{xx} = E\left(\frac{du_0}{dx} - z\frac{d^2w_0}{dx^2}\right) \tag{4.10}$$

Substitution of (4.10) into (4.7) and (4.8) yields

$$N_x = EA\frac{du_0}{dx} - E\frac{d^2w_0}{dx^2}\iint_A z\, dA \tag{4.11}$$

$$M_y = E\frac{du_0}{dx}\iint_A z\, dA - E\frac{d^2w_0}{dx^2}\iint_A z^2\, dA \tag{4.12}$$

Since the origin of the coordinates coincides with the centroid of the cross-section, we have

$$\iint_A z \, dA = 0$$

Thus, (4.11) and (4.12) reduce to

$$N_x = EA\frac{du_0}{dx} \qquad (4.13)$$

$$M_y = -EI_y\frac{d^2 w_0}{dx^2} \qquad (4.14)$$

where

$$I_y = \iint_A z^2 \, dA$$

is the moment of inertia of the cross-sectional area about the y-axis.

If no axial force is applied, i.e., $N_x = 0$, then $du_0/dx = 0$. From (4.5), this means that $\varepsilon_{xx} = 0$ along the x-axis (or, more precisely, in the x–y plane). Thus, the x-axis is the neutral axis and the x–y plane is the neutral plane.

After adopting the approximation of (4.9), we note that $\gamma_{xz} = 0$, and, as a result, the transverse shear stress τ_{xz} cannot be obtained from the shear strain (which is approximated to be zero). The resultant transverse shear force

$$V_z = \iint_A \tau_{xz} \, dA \qquad (4.15)$$

should be obtained from considering the equilibrium of a beam element as shown in Fig. 4.3.

The force equilibrium in the z-direction gives

$$\Delta V_z + p_z \, \Delta x = 0$$

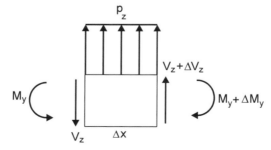

Figure 4.3   Equilibrium of a beam element.

Taking the limit $\Delta x \to 0$, we have

$$\frac{dV_z}{dx} = -p_z \tag{4.16}$$

The equilibrium of moments about the y-axis located at the left end of the beam element in Fig. 4.3 yields

$$\Delta M_y - p_z \Delta x \cdot \tfrac{1}{2}\Delta x - (V_z + \Delta V_z)\,\Delta x = 0$$

After taking $\Delta x \to 0$ (and thus $\Delta V_z \to 0$), we obtain

$$\frac{dM_y}{dx} = V_z \tag{4.17}$$

It is evident that the transverse shear force V_z can be derived from the bending moment using (4.17).

If the beam is subjected to a pure constant moment, then

$$V_z = \frac{dM_y}{dx} = 0$$

This is satisfied by setting $\tau_{xz} = 0$ [see (4.15)]. Evidently, the assumption $\gamma_{xz} = 0$ is exact in this case when moment is constant along the beam.

Substituting (4.14) into (4.17) and then into (4.16), we obtain

$$EI_y\frac{d^4w_0}{dx^4} = p_z \tag{4.18}$$

This is the **Bernoulli–Euler beam** (simple beam) **equation**.

In the absence of axial force, i.e., $N_x = 0$, we have $du_0/dx = 0$ from (4.13), and the bending strain reduces to

$$\varepsilon_{xx} = -z\frac{d^2w_0}{dx^2} \tag{4.19}$$

Using (4.14) and (4.19), we can write

$$\varepsilon_{xx} = \frac{M_y z}{EI_y} \tag{4.20}$$

and consequently,

$$\sigma_{xx} = \frac{M_y z}{I_y} \tag{4.21}$$

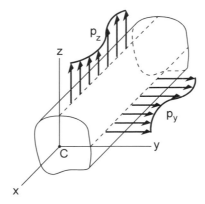

Figure 4.4   Beam with an arbitrary cross-section under bidirectional loading.

4.2 BIDIRECTIONAL BENDING

For beams with arbitrarily shaped cross-sections, we set up the coordinate system as shown in Fig. 4.4. Again, the x-axis is chosen to coincide with the centroidal axis. The external load is decomposed into p_y and p_z in the y and z directions, respectively. It is noted that the line loads must pass through the center of twist (shear center) if torsion is to be avoided. However, in the following development of the bending theory, the actual positions of application for line loads p_y and p_z need not be specified, although in Fig. 4.4 both p_y and p_z are shown to pass through the centroid of the cross-section.

Under such bidirectional bending, the longitudinal displacement is a function of x, y, and z. The approximate displacement expansions similar to (4.3) and (4.4) are given by

$$u = u_0(x) + z\psi_y(x) + y\psi_z(x) \tag{4.22a}$$

$$v = v_0(x) \tag{4.22b}$$

$$w = w_0(x) \tag{4.22c}$$

where ψ_y and ψ_z are rotations of the cross-section about the y and z axes, respectively. The *positive direction of ψ_y is the right-hand rotation about the positive y-axis, and ψ_z is about the negative z-axis.*

The corresponding strains are

$$\varepsilon_{xx} = \frac{\partial u}{\partial x} = \frac{du_0}{dx} + z\frac{d\psi_y}{dx} + y\frac{d\psi_z}{dx} \tag{4.23a}$$

$$\gamma_{xy} = \frac{\partial v}{\partial x} + \frac{\partial u}{\partial y} = \frac{dv_0}{dx} + \psi_z \tag{4.23b}$$

$$\gamma_{xz} = \frac{\partial w}{\partial x} + \frac{\partial u}{\partial z} = \frac{dw_0}{dx} + \psi_y \qquad (4.23c)$$

Again, the simplifying assumption $\gamma_{xy} = \gamma_{xz} = 0$ yields the relations

$$\psi_z = -\frac{dv_0}{dx}$$

$$\psi_y = -\frac{dw_0}{dx}$$

which are substituted into (4.23a) to obtain

$$\varepsilon_{xx} = \frac{du_0}{dx} - y\frac{d^2v_0}{dx^2} - z\frac{d^2w_0}{dx^2} \qquad (4.24)$$

Using the argument that $du_0/dx = 0$ if $N_x = 0$, the bending strain is reduced to

$$\varepsilon_{xx} = -y\frac{d^2v_0}{dx^2} - z\frac{d^2w_0}{dx^2} \qquad (4.25)$$

The bending moments about the y and z axes, respectively, are defined as

$$M_y = \iint_A z\sigma_{xx}\,dA = -E\iint_A \left(yz\frac{d^2v_0}{dx^2} + z^2\frac{d^2w_0}{dx^2} \right)dA$$

$$= -EI_{yz}\frac{d^2v_0}{dx^2} - EI_y\frac{d^2w_0}{dx^2} \qquad (4.26)$$

$$M_z = \iint_A y\sigma_{xx}\,dA = -EI_z\frac{d^2v_0}{dx^2} - EI_{yz}\frac{d^2w_0}{dx^2} \qquad (4.27)$$

where

$$I_y = \iint_A z^2\,dA \quad \text{moment of inertia about } y\text{-axis} \qquad (4.28a)$$

$$I_z = \iint_A y^2\,dA \quad \text{moment of inertia about } z\text{-axis} \qquad (4.28b)$$

$$I_{yz} = \iint_A yz\,dA \quad \text{product of inertia} \qquad (4.28c)$$

From the definitions of (4.26) and (4.27), the sign convention for M_y and M_z is determined as illustrated in Fig. 4.5.

Solving (4.26) and (4.27), we obtain

$$-E\frac{d^2v_0}{dx^2} = \frac{1}{I_yI_z - I_{yz}^2}(-I_{yz}M_y + I_yM_z)$$

$$-E\frac{d^2w_0}{dx^2} = \frac{1}{I_yI_z - I_{yz}^2}(I_zM_y - I_{yz}M_z)$$

Using (4.25), we write the bending stress as

$$\sigma_{xx} = E\varepsilon_{xx} = -yE\frac{d^2v_0}{dx^2} - zE\frac{d^2w_0}{dx^2}$$

$$= \frac{I_yM_z - I_{yz}M_y}{I_yI_z - I_{yz}^2}y + \frac{I_zM_y - I_{yz}M_z}{I_yI_z - I_{yz}^2}z \qquad (4.29)$$

The location of the **neutral axis** (neutral plane) along which $\sigma_{xx} = 0$ can be found from (4.29), i.e.,

$$\frac{I_yM_z - I_{yz}M_y}{I_yI_z - I_{yz}^2}y = -\frac{I_zM_y - I_{yz}M_z}{I_yI_z - I_{yz}^2}z$$

Defining the neutral plane by angle α as shown in Fig. 4.6, we have

$$\tan\alpha = -\frac{z}{y} = \frac{I_yM_z - I_{yz}M_y}{I_zM_y - I_{yz}M_z} \qquad (4.30)$$

Note that the positive direction of angle α is clockwise.

Since the distribution of the bending stress is linear over the cross-section, it is also linear with respect to the distance to the neutral axis. That is, the

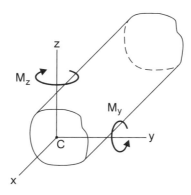

Figure 4.5   Sign convention of M_y and M_z.

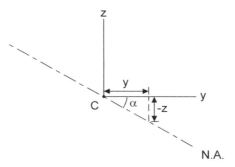

Figure 4.6 Neutral axis by angle α.

most distant location from the neutral axis experiences the greatest magnitude of bending stress. Further, the neutral axis divides the tensile bending stress field and the compressive stress field.

If the y- or z-axis is an axis of symmetry for the cross-section, then $I_{yz} = 0$ and (4.29) reduces to

$$\sigma_{xx} = \frac{M_z}{I_z}y + \frac{M_y}{I_y}z \tag{4.31}$$

Further, if $M_z = 0$, the bending stress becomes

$$\sigma_{xx} = \frac{M_y}{I_y}z \tag{4.32}$$

This is what was derived in Section 4.1 for symmetric sections.

If $I_{yz} \neq 0$ and $M_z = 0$, then from (4.29) we have

$$\sigma_{xx} = -\frac{I_{yz}M_y}{I_yI_z - I_{yz}^2}y + \frac{I_zM_y}{I_yI_z - I_{yz}^2}z \tag{4.33}$$

Therefore, for beams with an arbitrary cross-section under one-way bending, say $M_y \neq 0$ and $M_z = 0$, the simple beam bending stress formula (4.32) is not valid, and (4.33) must be used.

From the equilibrium considerations of a differential beam element, it is easy to derive the following relations for bidirectional bending by following the procedures used in deriving (4.16) and (4.17).

$$\frac{dV_z}{dx} = -p_z \tag{4.34a}$$

$$\frac{dM_y}{dx} = V_z \tag{4.34b}$$

$$\frac{dV_y}{dx} = -p_y \tag{4.34c}$$

$$\frac{dM_z}{dx} = V_y \tag{4.34d}$$

The relations in (4.34) are valid only for the sign convention adopted as illustrated in Fig. 4.3. Note that the positive directions of the shear force V_z and bending moment M_y acting on the positive cross-section (with an outward normal vector pointing in the positive x-direction) are opposite to those acting on the negative cross-section. Henceforth, the cross-sections and the accompanied positive shear forces will be shown to avoid confusion.

From (4.34) the equilibrium equations can be written as

$$\frac{d^2 M_y}{dx^2} = -p_z$$

$$\frac{d^2 M_z}{dx^2} = -p_y$$

Substituting (4.26) and (4.27) in the above equations, respectively, we obtain the displacement equilibrium equations for the beam theory:

$$E I_{yz} \frac{d^4 v_0}{dx^4} + E I_y \frac{d^4 w_0}{dx^4} = p_z$$

$$E I_z \frac{d^4 v_0}{dx^4} + E I_{yz} \frac{d^4 w_0}{dx^4} = p_y$$

For symmetric sections, $I_{yz} = 0$ and the above beam equations reduce to the form of (4.18).

Example 4.1   The cross-section of a single-cell box beam with four stringers is shown in Fig. 4.7. The contribution of the thin sheets to bending is assumed to be negligible. Thus, only the areas of the stringers are considered in the bending analysis. The areas of the stringers are $A_1 = 6 \times 10^{-4}$ m^2, $A_2 = 5 \times 10^{-4}$ m^2, $A_3 = A_4 = 4 \times 10^{-4}$ m^2.

First, the centroid of the effective cross-sectional areas (i.e., those of the stringers) must be determined. Denoting the coordinates of the stringers by $(\bar{y}_j, \bar{z}_i)$ with respect to the $\bar{y}$–$\bar{z}$ system, we have the coordinates of the centroid:

$$\bar{y}_c = \frac{\sum \bar{y}_i A_i}{\sum A_i} = \frac{0 A_1 + 0 A_2 + 0.5 A_3 + 0.5 A_4}{A_1 + A_2 + A_3 + A_4} = \frac{4 \times 10^{-4}}{19 \times 10^{-4}} = 0.21 \text{ m}$$

$$\bar{z}_c = \frac{\sum \bar{z}_i A_i}{\sum A_i} = \frac{0.4A_1 + 0A_2 + 0A_3 + 0.3A_4}{A_1 + A_2 + A_3 + A_4} = \frac{3.6}{19} = 0.19 \text{ m}$$

Thus, the location of the centroid is $(0.21, 0.19)$ in the $\bar{y}$–$\bar{z}$ system.

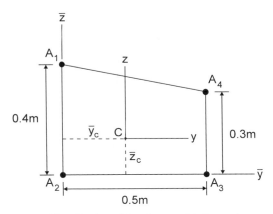

Figure 4.7   Single-cell box beam with four stringers.

This cross-section consisting of four stringers is not symmetric with respect to either the y- or the z-axis. Hence, the general bending equations must be used. The moments of inertia of the effective cross-sectional area of the box beam with respect to the coordinate system y–z are calculated according to (4.28). Denoting the coordinates of each stringer by (y_i, z_i) with respect to the y–z system, we have

$$I_y = \sum A_i z_i^2$$
$$= A_1(0.4 - 0.19)^2 + (A_2 + A_3)(0.19)^2 + A_4(0.3 - 0.19)^2$$
$$= 0.63 \times 10^{-4} \text{ m}^4$$
$$I_z = \sum A_i y_i^2$$
$$= (A_1 + A_2)(0.21)^2 + (A_3 + A_4)(0.29)^2 = 1.16 \times 10^{-4} \text{ m}^4$$
$$I_{yz} = \sum A_i y_i z_i$$
$$= A_1(-0.21)(0.21) + A_2(-0.21)(-0.19) + A_3(0.29)(-0.19)$$
$$+ A_4(0.29)(0.11)$$
$$= -0.15 \times 10^{-4} \text{ m}^4$$

Consider the loading $M_z = 0$ and $M_y \neq 0$. The neutral plane is given by

$$\tan \alpha = -\frac{I_{yz}M_y}{I_z M_y} = \frac{0.15 \times 10^{-4}}{1.16 \times 10^{-4}} = 0.13$$

which yields $\alpha = 7°$ measured clockwise from the y-axis.

The bending stresses in the stringers can be calculated easily using (4.29), if M_y and M_z are given.

Example 4.2 A beam of the thin-walled Z-section shown in Fig. 4.8 is subjected to a positive bending moment M_y. Find the distribution of bending stresses.

In this case, the centroid is easy to locate to be at the midpoint of the vertical web as shown in Fig. 4.8.

The moment of inertia of the cross-section is the sum of the moments of inertia of the three rectangular sections of the web and two flanges. For each section it is most convenient to calculate I_y, I_z, and I_{yz} using the parallel axis method. For example, consider the upper flange 3–4. Let the y' and z' axes be local coordinates with the origin at the centroid C' of the upper flange. According to the parallel axis theorem, the moment of inertia I_y of the cross-section of the upper flange is

$$I_y = A_{34} \left(\frac{h}{2}\right)^2 + I_{y'} \tag{a}$$

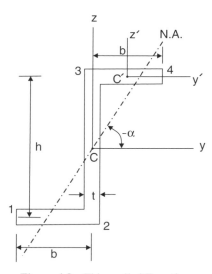

Figure 4.8 Thin-walled Z-section.

in which A_{34} is the cross-sectional area of the upper flange and $I_{y'}$ is the moment of inertia about y'-axis. For the upper flange, we have

$$I_{y'} = \frac{bt^3}{12}$$

Similarly, the moment of inertia about the z-axis can be obtained in a similar manner.

The product of inertia of the upper flange can be written as

$$I_{yz} = A_{34}\frac{h}{2}\frac{b}{2} + I_{y'z'} \tag{b}$$

and $I_{y'z'} = 0$ because of symmetry with respect to the local coordinate system.

For the entire Z-section, we obtain

$$I_y = 2bt\left(\frac{h}{2}\right)^2 + \frac{2bt^3}{12} + \frac{th^3}{12} \tag{c}$$

$$I_z = 2bt\left(\frac{b}{2}\right)^2 + \frac{2tb^3}{12} + \frac{ht^3}{12} \tag{d}$$

$$I_{yz} = bt\frac{b}{2}\frac{h}{2} + bt\left(-\frac{b}{2}\right)\left(-\frac{h}{2}\right) = \frac{b^2ht}{2} \tag{e}$$

For thin-walled sections, the terms with t^3 are small and are neglected in the following calculations.

The orientation of the neutral axis can be calculated using (4.30) with $M_z = 0$. Then

$$\tan\alpha = -\frac{I_{yz}}{I_z} = -\frac{3h}{4b} \tag{f}$$

Thus, $\alpha = -\tan^{-1}(3h/4b)$. Note that a negative value of α means that the neutral axis is oriented at a counterclockwise rotation of an angle $-\alpha$ from the y-axis as shown in Fig. 4.8. Since the bending stress distribution is linear with respect to the distance (positive above and negative below the neutral axis) from the neutral axis, the distribution on the Z-section must be antisymmetric with respect to the neutral axis.

Let us consider the case with $h = 2b$, which leads to $\alpha = -\tan^{-1}1.5 = -56.3°$. Note that in this case, points 3 and 4 on the upper flange are on opposite sides of the neutral axis, and the corresponding bending stresses must be of opposite signs. At the free edge of the upper flange (point 4 in Fig. 4.8) the bending stress is calculated using (4.33) together with $y = h/2$,

$z = h/2$. We have, for a positive bending moment,

$$\sigma_{xx} = -\frac{1.713 M_y}{th^2} \qquad \text{(compression)}$$

Similarly, at point 3 ($y = 0$, $z = h/2$), we obtain

$$\sigma_{xx} = \frac{3.426 M_y}{th^2} \qquad \text{(tension)}$$

4.3   TRANSVERSE SHEAR STRESS DUE TO TRANSVERSE FORCE IN SYMMETRIC SECTIONS

In deriving the Bernoulli–Euler beam equation, the transverse shear strain γ_{xz} was neglected while the transverse shear stress τ_{xz} (and, thus, the transverse shear force V_z) was kept in the equilibrium equation. Such contradictions are often found in simplified structural theories. The assumption $\gamma_{xz} = 0$ is quite good for slender beams (i.e., the depth is small compared with the span). In fact, it is exact if the loading is a pure bending moment. However, for short beams under transverse loads, significant shear stress (strain) may result.

The exact distribution of τ_{xz} on the cross-section of a beam subjected to transverse forces is generally not easy to analyze. An exception is the narrow rectangular section as shown in Fig. 4.9. If $h \gg b$, the plane stress assumption adopted in the derivation of the simple beam theory is valid. In other words, τ_{xz} can be assumed to be uniform across the width of the section. Otherwise, τ_{xz} is a function of y.

4.3.1   Narrow Rectangular Cross-Section

Consider a beam with a narrow rectangular cross-section as shown in Fig. 4.9. The resultant transverse shear force V_z is

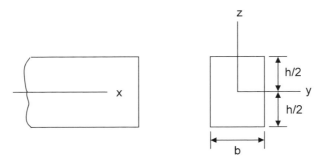

Figure 4.9   Narrow rectangular section.

$$V_z = b \int_{-h/2}^{h/2} \tau_{xz} \, dz$$

This definition alone is not sufficient to recover the distribution of τ_{xz} in the vertical (z) direction. We resort to the equilibrium equation for a state of plane stress parallel to the x–z plane [see (2.112)],

$$\frac{\partial \sigma_{xx}}{\partial x} + \frac{\partial \tau_{xz}}{\partial z} = 0 \tag{4.35}$$

Substituting (4.21) into (4.35), we obtain

$$\frac{z}{I_y} \frac{\partial M_y}{\partial x} + \frac{\partial \tau_{xz}}{\partial z} = 0 \tag{4.36}$$

Using the relation

$$\frac{\partial M_y}{\partial x} = V_z$$

in (4.36) we have

$$\frac{\partial \tau_{xz}}{\partial z} = -\frac{z V_z}{I_y}$$

Integrating the above equation from $z = -h/2$ to z, we obtain

$$\tau_{xz}(z) - \tau_{xz}\left(-\frac{h}{2}\right) = -\frac{V_z}{2I_y}\left(z^2 - \frac{h^2}{4}\right) \tag{4.37}$$

Since the shear stress vanishes at the top and bottom faces, i.e.,

$$\tau_{xz}\left(\pm\frac{h}{2}\right) = 0$$

(4.37) reduces to

$$\tau_{xz} = \frac{V_z c^2}{2I_y}\left(1 - \frac{z^2}{c^2}\right) \tag{4.38}$$

where

$$c = \frac{h}{2}$$

From (4.38), it is evident that τ_{xy} has a parabolic distribution over z, and the maximum value which occurs at $z = 0$ is

$$(\tau_{xz})_{\text{max}} = \frac{V_z c^2}{2I_y}$$

4.3.2   General Symmetric Sections

For uniform beams with general symmetric (with respect to the z-axis) cross-sections, the simple beam results are valid, i.e.,

$$\varepsilon_{xx} = \frac{M_y z}{E I_y} \tag{4.39}$$

$$\sigma_{xx} = \frac{M_y z}{I_y} \tag{4.40}$$

$$V_z = \frac{d M_y}{dx} \tag{4.41}$$

However, the transverse shear stress τ_{xz} distribution over the cross-section is difficult to analyze. For symmetrical sections under a transverse shear force V_z, the only thing we know is that the distribution of τ_{xz} is symmetrical with respect to the z-axis.

Since the variation of τ_{xz} in the y-direction is unknown, it is more convenient to consider the transverse shear flow q_z defined as

$$q_z(z) = \int \tau_{xz}\, dy \tag{4.42}$$

If τ_{xz} is uniform across the width t, then

$$q_z = t \tau_{xz} \tag{4.43}$$

If τ_{xz} is not uniform in the y-direction, the average value is introduced as

$$\bar{\tau}_{xz} = \frac{q_z}{t} \tag{4.44}$$

Here, the positive direction of q_z is taken to coincide with that of τ_{xz}.

The transverse shear flow q_z can be determined from the equilibrium of a differential beam element as shown in Fig. 4.10. In Fig. 4.10a, the side view of the beam element of length Δx is shown with the bending stresses acting on the two neighboring cross-sections. Consider the free body of the beam element above the B–B plane (i.e., $z \geq z_1$) as shown in Fig. 4.10b. The shear flow q_z on the cross-section at $z = z_1$ is equal to the shear force acting on

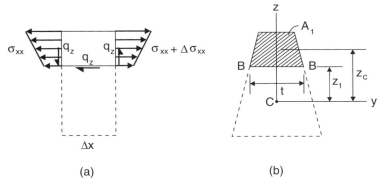

Figure 4.10 Differential beam element with bending stresses: (a) longitudinal section; (b) cross-section.

the bottom face of the free body as depicted in Fig. 4.10a. For the free body above $z = z_1$, equilibrium equation $\sum F_x = 0$ is given by

$$\iint_{A_1} \Delta\sigma_{xx}\, dA = q_z\, \Delta x \tag{4.45}$$

where A_1 is the cross-sectional area above $z = z_1$. Dividing both sides of (4.45) by Δx and taking the limit $\Delta x \to 0$, we have

$$\iint_{A_1} \frac{d\sigma_{xx}}{dx}\, dA = q_z \tag{4.46}$$

By using (4.40) in (4.46) we obtain the transverse shear flow as

$$q_z = \iint_{A_1} \frac{dM_y}{dx}\frac{z}{I_y}\, dA$$

$$= \frac{dM_y}{dx}\frac{1}{I_y} \iint_{A_1} z\, dA$$

$$= \frac{V_z Q}{I_y} \tag{4.47}$$

where

$$Q = \iint_{A_1} z\, dA \tag{4.48}$$

is the first moment of the area A_1. If the centroid of A_1 is at $z = z_c$, then Q can also be expressed as

$$Q = A_1 z_c \tag{4.49}$$

4.3.3 Thin-Walled Sections

Consider the wide-flange beam shown in Fig. 4.11a. The transverse shear stress is usually represented by the average value $\bar{\tau}_{xz}$ as shown in Fig. 4.11b. A jump in $\bar{\tau}_{xz}$ is noted at plane $CDEF$ due to the sudden change of width. If the transverse shear flow q_z is plotted as shown in Fig. 4.11c, then no such jump exists.

In the free-hung portion of the wide flange (e.g., the portions CD and EF), the actual transverse shear stress τ_{xz} is much smaller than the average value $\bar{\tau}_{xz}$. It is noted that τ_{xz} must vanish along AB, CD, and EF. If the thickness of the flange is small, then τ_{xz} cannot build up significantly except for the portion connected to the vertical web. A more accurate distribution of τ_{xz} along $C'D'E'F'$ is depicted in Fig. 4.12.

From Fig. 4.11, it is seen that for wide-flange beams, the transverse shear stress is small in the flange and that the web carries the majority of the transverse shear load. An approximate model for such a wide-flange beam is obtained by lumping the total area of the flange into a concentrated area

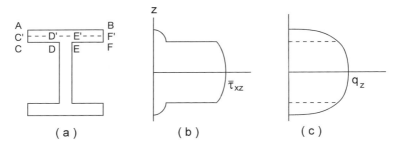

Figure 4.11   (a) Wide-flange beam; (b) distribution of $\bar{\tau}_{xz}$; (c) distribution of shear flow q_z.

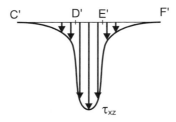

Figure 4.12   Distribution of τ_{xz} in a wide-flange beam.

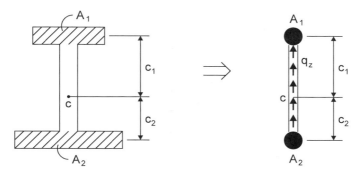

Figure 4.13   Concentration area for a wide flange.

as shown in Fig. 4.13. In addition, we may assume that the web does not contribute to resisting bending. Thus, for the web section we have

$$Q = A_1 c_1$$

which remains unchanged with location. As a result q_z is constant along the web.

In aircraft structures, stringers are often used to provide bending stiffness, and thin webs are used to carry shear flows. To maximize the bending capacity of the structure, we place stringers at the greatest distance from the neutral axis. The thin web is usually assumed ineffective in bending. Consequently, the shear flow in the web between two adjacent stringers is constant.

It should be noted that although the transverse shear stress τ_{xz} is very small in the flanges, the in-plane shear stress τ_{xy} is significantly large, as discussed in Chapter 5. However, the existence of the in-plane shear stress does not affect the simplified calculation above of the transverse shear flow in the web.

4.3.4   Shear Deformation in Thin-Walled Sections

In developing the simple beam theory in Sections 4.1 and 4.2, the transverse shear strains are neglected, leading to the well-known assumptions in the simple beam theory that plane sections remain plane and normal to the neutral axis after deformation. However, this simplification may lead to substantial errors in estimating the deflection of thin-walled beam members unless they are long or they are under pure bending moments.

Consider the thin-walled beam loaded as shown in Fig. 4.14a. Note that at the free end no shear stress is applied. We assume that the web can only take shearing stresses, and bending is taken by the two stringers. The bending

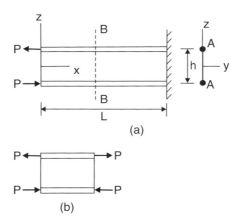

Figure 4.14   Stringer–web beam under pure bending.

moment is $M = Ph$ and is uniform over the entire length of the beam. From the free body diagram in Fig. 4.14b, at any section B–B, there is no transverse shear stress (and thus no transverse shear strain) in the web based on the equilibrium condition $\sum F_z = 0$. Hence, we conclude that the axial forces in the stringers also remain constant over the length of the beam. The assumptions adopted by the simple beam theory are thus valid in this case.

Consider now the beam of Fig. 4.14 subjected to a shear load (a constant shear flow q_0) at the free end as shown in Fig. 4.15a. From the force and moment balance conditions of the free body shown in Fig. 4.15b, we have a constant transverse shear flow $q_z = q_0$ and linearly varying axial force $P = q_0 h x / h = q_0 x$ over the beam length. The corresponding moment along the beam is $M = Ph = q_0 h x$.

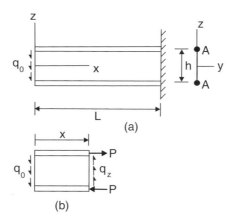

Figure 4.15   Stringer–web beam subjected to a transverse shear force.

The deflection of the beam consists of two parts; one part results from the bending moment and the other from the transverse shear deformation. The deflection due to bending moment is obtained from the simple beam equation (4.14):

$$\frac{d^2 w_0}{dx^2} = -\frac{M_y}{E I_y} = -\frac{q_0 h}{E I_y} x \tag{4.50}$$

Integrating (4.50) twice, we obtain

$$w_0 = -\frac{q_0 h}{6 E I_y} x^3 + C_1 x + C_2 \tag{4.51}$$

where C_1 and C_2 are arbitrary constants. From the boundary conditions

$$w_0 = 0 \quad \text{and} \quad \frac{d w_0}{dx} = 0 \qquad \text{at } x = L$$

C_1 and C_2 are determined to be

$$C_1 = \frac{q_0 h L^2}{2 E I_y}, \qquad C_2 = -\frac{q_0 h L^3}{3 E I_y}$$

Then the deflection at the free end ($x = 0$) due to bending deformation is obtained from (4.51) as

$$w_0^B = -\frac{q_0 h L^3}{3 E I_y} \tag{4.52}$$

The shear deformation in the web contributes to the beam deflection in addition to the bending moment. The constant shear strain in the web is

$$\gamma_{xz} = \frac{\tau_{xz}}{G} = \frac{q_0}{t G} \tag{4.53}$$

in which t is the thickness of the web. Recall that in deriving the simple beam theory, we set $\gamma_{xz} = 0$. Now we relax the condition and allow the transverse shear strain to occur and use the relation

$$\gamma_{xz} = \frac{\partial w}{\partial x} + \frac{\partial u}{\partial z} \tag{4.54}$$

to calculate the transverse displacement. At the fixed end ($x = L$), $\partial u / \partial z = 0$, which is true over the entire beam length because γ_{xz} is constant. Since γ_{xz} is uniform in the web, we have

$$\gamma_{xz} = \frac{\partial w}{\partial x} = \frac{d w_0}{dx} = \frac{q_0}{t G}$$

along the entire beam. Subsequently, we obtain by integration

$$w_0 = \frac{q_0}{tG}x + C$$

The integration constant C can be determined by the boundary condition $w_0 = 0$ at $x = L$. The result is

$$C = -\frac{q_0 L}{tG} \tag{4.55}$$

Thus, the deflection at the free end ($x = 0$) due to shear deformation in the web is

$$w_0^s = -\frac{q_0 L}{tG} \tag{4.56}$$

Consider the ratio

$$\frac{w_0^S}{w_0^B} = \frac{3EI_y}{GAL^2} \tag{4.57}$$

where $A = th$ is the cross-sectional area of the web. It is evident that a long beam ($L \gg h$) with either a high shear rigidity (GA) or small bending rigidity (EI_y) would fit the description of the simple beam theory.

4.4 TIMOSHENKO BEAM THEORY

A beam theory that accounts for transverse shear deformation can be developed following the procedure presented in Section 4.1 for the simple beam theory. The displacement expansions of (4.3) and (4.4) are adopted, but the zero transverse shear strain condition (4.9) is not imposed. If, again, we assumed that the axial force N_x is absent, then only transverse deflection w_0 and rotation of cross-section ψ_y are kept in the formulation. The only nontrivial strains are

$$\varepsilon_{xx} = z\frac{d\psi_y}{dx} \tag{4.58}$$

$$\gamma_{xz} = \frac{dw_0}{dx} + \psi_y \tag{4.59}$$

Since the transverse shear strain γ_{xz} exists, the transverse shear stress τ_{xz} can be calculated directly from γ_{xz} using the stress-strain relations. The resultant shear force acting on the cross-section can be calculated directly from the shear strain as

$$
V_z = \iint\limits_{A} \tau_{xz}\, dA = \iint\limits_{A} G\gamma_{xz}\, dA = \iint\limits_{A} G\left(\frac{dw_0}{dx} + \psi_y\right) dA
$$

$$
= GA\left(\frac{dw_0}{dx} + \psi_y\right) \tag{4.60}
$$

where A is the area of the cross-section that carries transverse shear loads.

The bending moment is defined by (4.8). Using $\sigma_{xx} = E\varepsilon_{xx}$ together with (4.58) in (4.8), we have

$$
M_y = EI_y \frac{d\psi_y}{dx} \tag{4.61}
$$

The equilibrium conditions (4.16) and (4.17), which are obtained from the balance of vertical forces and balance of moments of a differential beam element (see Fig. 4.3), remain valid. Substituting (4.60) and (4.61) in (4.17) and (4.16), we obtain

$$
EI_y \frac{d^2\psi_y}{dx^2} - GA\left(\frac{dw_0}{dx} + \psi_y\right) = 0 \tag{4.62}
$$

$$
GA\left(\frac{d^2w_0}{dx^2} + \frac{d\psi_y}{dx}\right) + p_z = 0 \tag{4.63}
$$

These are the two equilibrium equations of the **Timoshenko beam theory**. The two equilibrium equations may be combined into one by eliminating ψ_y. This can be performed by first solving (4.63) for $d\psi_y/dx$ and then substituting it in (4.62) after differentiating it once. The result is

$$
EI_y \frac{d^4w_0}{dx^4} = p_z - \frac{EI_y}{GZ}\frac{d^2p_z}{dx^2} \tag{4.64}
$$

The boundary conditions at the ends of a Timoshenko beam are specified as follows.

- For a hinged end, $w_0 = 0$, and M_y is prescribed.
- For a clamped end, $w_0 = \psi_y = 0$.
- For a free end, V_z and M_y are prescribed.

Note that at a clamped end, the rotation of the cross-section vanishes while the slope of deflection dw_0/dx is not zero unless the transverse shear strain vanishes at that location.

A Timoshenko beam theory for bidirectional bending can be derived in a similar manner.

Example 4.3   The problem of the two-stringer beam shown in Fig. 4.15a may be solved by using the Timoshenko beam theory to achieve a more accurate deflection. Two solution procedures may be used. The first one is to solve the combined equilibrium equation (4.64). Since $p_z = 0$, (4.64) reduces to

$$EI_y \frac{d^4 w_0}{dx^4} = 0$$

for which the general solution is readily obtained as

$$w_0 = C_1 x^3 + C_2 x^2 + C_3 x + C_4$$

From (4.63) we have

$$\frac{d\psi_y}{dx} = -\frac{d^2 w_0}{dx^2}$$

which is then substituted in (4.62) to yield

$$\psi_y = -\frac{dw_0}{dx} - \frac{EI_y}{GA} \frac{d^3 w_0}{dx^3}$$

Thus, the function ψ_y also contains the same integration constants, C_1, C_2, C_3, and C_4. Boundary conditions are used to determine these four constants.

An alternative approach is to take advantage of the fact that the shear loading at the free end produces a constant shear force along the beam, i.e.,

$$V_z = q_0 h = GA \left(\frac{dw_0}{dx} + \psi_y \right) = \text{constant} \qquad \text{(a)}$$

Substitution of (a) in (4.62) yields

$$\frac{d^2 \psi_y}{dx^2} = \frac{q_0 h}{EI_y}$$

and after integration of the differential equation above,

$$\psi_y = \frac{q_0 h}{2EI_y} x^2 + B_1 x + B_2 \qquad \text{(b)}$$

where B_1 and B_2 are integration constants.

From (4.63), we have

$$\frac{d^2 w_0}{dx^2} + \frac{d\psi_y}{dx} = 0$$

Thus,

$$\frac{d^2 w_0}{dx^2} = -\frac{d\psi_y}{dx} = -\frac{q_0 h}{E I_y} x - B_1 \tag{c}$$

Integrate (c) twice to obtain

$$w_0 = -\frac{q_0 h}{6 E I_y} x^3 - \frac{1}{2} B_1 x^2 + B_3 x + B_4 \tag{d}$$

The boundary conditions are

$$M_y = E I_y \frac{d\psi_y}{dx} = 0 \qquad \text{at } x = 0$$

$$V_z = q_0 h = G A \left(\frac{d w_0}{dx} + \psi_y \right) \qquad \text{at } x = 0 \text{ (this condition has already been used)}$$

$$w_0 = 0 \qquad \text{at } x = L$$

$$\psi_y = 0 \qquad \text{at } x = L$$

Solving the four equations above with the general solutions given by (b) and (d), we obtain

$$B_1 = 0$$

$$B_2 = -\frac{q_0 h L^2}{E I_y}$$

$$B_3 = \frac{q_0 h L^2}{2 E I_y} + \frac{q_0 h}{G A}$$

$$B_4 = \frac{q_0 h L^3}{6 E I_y} - B_3 L$$

and thus, the deflection curve

$$w_0 = -\frac{q_0 h}{6 E I_y} (x^3 - L^3) + B_3 (x - L) \tag{e}$$

The maximum deflection occurs at the free end ($x = 0$):

$$w_0(0) = -\frac{q_0 h L^3}{3 E I_y} - \frac{q_0 h L}{G A} \tag{f}$$

It is evident that the first term in the solution above is the deflection associated with bending given by (4.52) and the second term is associated with transverse shear deformation given by (4.56).

In the derivation of the Timoshenko beam theory, two terms in the displacement expansion (4.1) are kept, resulting in a constant transverse shear strain over the entire cross-section [see (4.6)]. To compensate for such an oversimplification, it is customary to introduce a shear correction factor k to modify the shear rigidity GA of the cross-section into kGA. For a rectangular solid cross-section, the shear correction factor is often taken as $k = 5/6$ for static cases and $k = \pi^2/12$ for dynamic cases. For thin-walled sections with stiffeners, the wall that bears the shear load is often assumed to take no bending and the shear stress and strain in the thin-walled section become constant between two adjacent stiffeners as discussed in Example 4.3. Then setting $k = 1$ is consistent with the constant shear assumption. It should be noted that for curved or inclined walls, the shear flow direction follows that of the centerline of the wall. This type of shear flow is discussed in Chapter 5. Since the Timoshenko beam theory assumes that the resultant shear force acts in the z-direction, only the z-component of the shear flow should be taken into account in calculating the shear force V_z. Using the concept provided by (3.49), this effect of inclined shear flow can be accounted for by redefining the area A in the shear rigidity GA as the projection of the area of the inclined cross-section on the z-axis. For example, for both sections shown in Fig. 4.16, $A = th$, rather than the actual cross-sectional area, should be used in the Timoshenko beam equations.

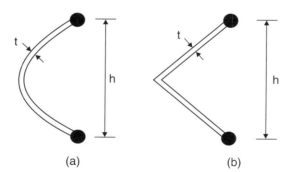

(a)   (b)

Figure 4.16   Curved and inclined thin-walled sections with stiffeners.

4.5 SHEAR LAG

Shear lag is a phenomenon of load distribution. Saint-Venant's principle is a good example of shear lag. Consider the stringer-sheet panel of Fig. 3.2a

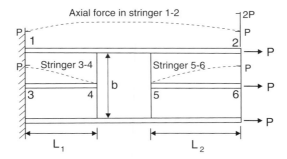

Figure 4.17   Cutout in a stringer sheet panel.

with three stringers loaded at the left end. At a distance from the loading end, the loads in the stringers all approach zero. The analysis presented in Section 3.1 indicates that this load redistribution is accomplished by the load transfer in the shear panels through a distance x_0. The exponential decay of the load in the two outside stringers is given by (3.8). The decay index λ given by (3.9) depends on the axial rigidity EA of the stringer and shear rigidity Ght of the sheet. A very rigid sheet makes the shear lag region small, while rigid stringers require a long distance to accomplish load redistribution.

In aircraft structures there are many locations where shear lags occur. The most notable ones are (1) cutouts where some stringers are discontinuous, and (2) sites of load application, which can be illustrated by the example of Fig. 3.2a. In essence, the mechanics of shear lag in the two situations above are basically the same.

Figure 4.17 shows a three stringer–web panel with a cutout. The loads carried by the stringers at the free end are the same. Due to the discontinuity, the load in stringer 5–6 drops to zero at the cutout, while the load in each of the two side stringers increases to $1.5P$ in the region. Load redistribution takes place again through shear lag beyond the cutout; the load in stringer 3–4 increases from zero and approaches P if length L_1 is large enough. The size of the shear lag zone near a cutout depends on the geometry of the structure and material properties of the stringer and the web. If stringers and webs are made of the same material, the shear lag zone is often taken as an approximation to be three times the cutout size b (see Fig. 4.17).

Example 4.4   A stringer-sheet cantilever I-beam is built in at one end and loaded by a moment at the free end as shown in Fig. 4.18a. Through the shear lag action, the axial stresses in the stringers at a distance from the loading end can be calculated using the beam theory. Find the distance of the shear lag.

Since there is no transverse shear loading, the moment produced by the pair of forces of opposite directions applied at the mid stringers is a pure

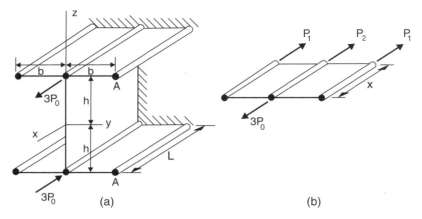

Figure 4.18   Shear lag in a stringer-sheet I-beam under pure bending.

bending and no shear stress is present in the vertical web. As a result, the top and bottom flanges can be analyzed separately. To simplify the analysis, we assume that sheets are ineffective in carrying normal stresses. The procedure of the analysis follows that used in Section 4.1 in the discussion of Saint-Venant's principle.

If the simple beam theory is employed to solve this bending problem, we find that the axial force in the three upper stringers is equal to P_0. This contradicts the actual loading condition at the free end of the structure.

Consider the free body diagram for the upper flange panel (see Fig. 4.18b). The balance of forces in the x-direction yields

$$P_2 = 3P_0 - 2P_1 \qquad (a)$$

Using a free body diagram similar to Fig. 3.2c, equations in the form of (3.2) and (3.4) are obtained, i.e.,

$$\tau = \frac{1}{t}\frac{dP_1}{dx} \qquad (b)$$

$$\frac{d\tau}{dx} = \frac{G}{Eb}\left(\frac{P_1}{A} - \frac{P_2}{A}\right) = \frac{3G}{EAb}(P_1 - P_0) \qquad (c)$$

where τ is the shear stress in the flange. Substitution of (b) in (c) leads to

$$\frac{d^2 P_1}{dx^2} - \lambda^2 P_1 = -\lambda^2 P_0 \qquad (d)$$

where

$$\lambda^2 = \frac{3Gt}{EAb}$$

The general solution to the second-order differential equation is readily obtained as

$$P_1 = Ce^{-\lambda x} + De^{\lambda x} + P_0 \tag{e}$$

in which the arbitrary constants C and D are to be determined by boundary conditions. At $x = 0$, $P_1 = 0$. Then

$$D = -C - P_0 \tag{f}$$

At $x = L$ (the built-in end), $\tau = 0$. In view of (b), this condition is equivalent to

$$\frac{dP_1}{dx} = 0$$

which leads to

$$-C\lambda e^{-2\lambda L} + D\lambda e^{\lambda L} = 0 \tag{g}$$

Solving (f) and (g), we obtain

$$C = -\frac{P_0}{1 + e^{-2\lambda L}} \quad \text{and} \quad D = -\frac{P_0}{1 + e^{2\lambda L}}$$

The force in the side stringer is then given by

$$\frac{P_1}{P_0} = 1 - \left(\frac{e^{-\lambda x}}{1 + e^{-2\lambda L}} + \frac{e^{\lambda x}}{1 + e^{2\lambda L}} \right) \tag{h}$$

At $x = L$,

$$\frac{P_1}{P_0} = 1 - \left(\frac{e^{-\lambda L}}{1 + e^{-2\lambda L}} + \frac{e^{\lambda L}}{1 + e^{2\lambda L}} \right) = 1 - \frac{2}{e^{\lambda L} + e^{-\lambda L}} \tag{i}$$

It is evident that as $L \to \infty$, $P_1 \to P_0$ and $P_2 \to P_0$. Then the beam solution is obtained.

It is of interest to know the value of L at which $P_1 = 0.99 P_0$. This value can be obtained by solving the following equation derived from (i):

$$0.99 = 1 - \frac{2}{e^{\lambda L} + e^{-\lambda L}}$$

The result is

$$\cosh \lambda L = 100 \quad \text{or} \quad \lambda L = 5.5 \tag{j}$$

For a material with $E = 70$ GPa, $G = 27$ GPa, $t = 2$ mm, $b = 200$ mm, and $A = 100$ mm^2, we have $\lambda = 0.0108$ mm^{-1}, and then the value of L at which $P_1 = 0.99P_0$ is $L = 490$ mm. This length is about 2.5 times the width b of the flange. At this location, the bending stresses in the stringers obtained from the simple beam theory are accurate.

PROBLEMS

4.1 A uniform beam of a thin-walled angle section as shown in Fig. 4.19 is subjected to the bending M_y ($M_z = 0$). Find the neutral axis and bending stress distribution over the cross-section.

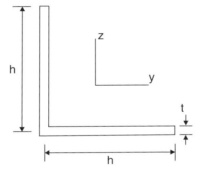

Figure 4.19   Thin-walled angle section.

4.2 Rotate the angle section of Fig. 4.19 counterclockwise for $45°$. Find the neutral axis and the maximum bending stress. Compare the load capacity with that of the original section given by Fig. 4.19.

4.3 The stringer–web sections shown in Figs. 4.20, 4.21, and 4.22 are subjected to the shear force $V_z \neq 0$, while $V_y = 0$. Find the bending

Figure 4.20   Stringer–web section.

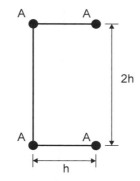

Figure 4.21 Stringer–web section.

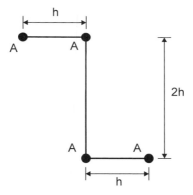

Figure 4.22 Stringer–web section.

stresses in the stringers for the same bending moment M_y. Which section is most effective in bending?

4.4 Compare the bending capabilities of the two sections of Figs. 4.21 and 4.22 if $M_y = 0$, $M_z \neq 0$.

4.5 Figure 4.23 shows the cross-section of a four-stringer box beam. Assume that the thin walls are ineffective in bending and the applied bending moments are

$$M_y = -500,000 \text{ N·cm}$$
$$M_z = 200,000 \text{ N·cm}$$

Find the bending stresses in all stringers.

4.6 Find the neutral axis in the thin-walled section of Fig. 4.23 for the loading given in Problem 4.5.

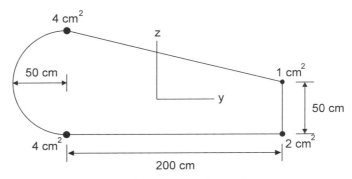

Figure 4.23 Thin-walled section.

4.7 Find the bending stresses in the stringers at the fixed end of the box beam loaded as shown in Fig. 4.24. Assume that the thin sheets are negligible in bending. Find the neutral axis.

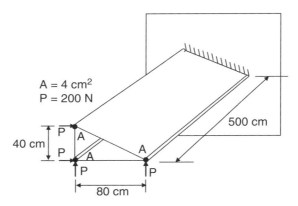

Figure 4.24 Loaded box beam.

4.8 Find the deflection of the box beam of Fig. 4.24 using the simple beam theory.

4.9 Find the bending stresses in the stringers of the box beam in Fig. 4.24 for the bending moments given in Problem 4.5.

4.10 A cantilever beam of a solid rectangular cross-section is loaded as shown in Fig. 4.25. Assume that the material is isotropic. Find the deflections of the beam using the simple beam theory and Timoshenko beam theory, respectively. Plot the ratio of the maximum deflections of the two solutions (at the free end) versus L/h. Use the shear correction factor $k = \frac{5}{6}$.

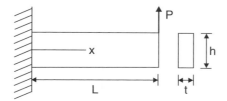

Figure 4.25   Cantilever beam subjected to a shear force P.

4.11 A thin-walled beam of length 5 m long with one end built into a rigid wall and the other end is subjected to a shear force $V_z = 5000$ N. The cross-section is given by Fig. 4.21 with $h = 0.2$ m and the wall thickness $= 0.002$ m. The material is aluminum 2024-T3 with $E = 70$ GPa, $G = 27$ GPa, and cross-sectional area of the stringer $A = 80$ mm^2. Assume that thin walls carry only shear stresses. Find the deflections at the free end using the simple beam theory and the Timoshenko beam theory, respectively. Compare the transverse shear stresses in the vertical web obtained from the two theories.

4.12 A 2024-T3 aluminum box beam with a thin-walled section is shown in Fig. 4.26. Assume that thin walls (thickness t) are ineffective in bending. Find the deflections at the free end using the simple beam theory for shear loads $V_z = 5000$ N and $V_y = 5000$ N separately. Solve the same problem using the Timoshenko beam theory. In which

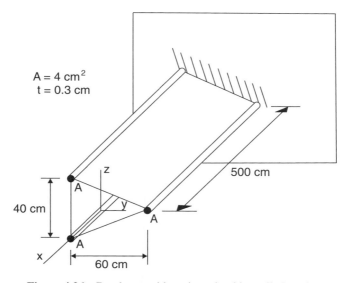

Figure 4.26   Box beam with a triangular thin-walled section.

loading case is the simple beam theory more accurate in predicting the deflection? Explain.

4.13 Consider the structure with a cutout as shown in Fig. 4.17. Find the axial force distribution in stringers 3–4 and 5–6. Assume that both stringers and webs have the same material properties of $E = 70$ GPa and $G = 27$ GPa. Also assume that $b = 200$ mm, the thickness of the web $t = 2$ mm, and the cross-sectional area of the stringer $A = 64$ mm^2. *Hint:* The zero-stress condition in the web at the cutout cannot be enforced because of the simplified assumption that shear stress and strain are uniform across the width of the web. Use the known condition that the force in the side stringers is $1.5P$ at the cutout.

5

FLEXURAL SHEAR FLOW IN THIN-WALLED SECTIONS

Shear flows in thin-walled sections can be generated by torques or transverse shear forces. The presence of shear strains along the wall gives rise to primary warping (in the longitudinal direction), and end constraints of the structure have significant effects on its structural behavior as discussed in Chapters 3 and 4. In this chapter, the shear flow produced by combined torsional and transverse loads is studied. To avoid mathematical complexities, the end-constraining effect is neglected in the analysis presented in this chapter.

5.1 FLEXURAL SHEAR FLOW IN OPEN THIN-WALLED SECTIONS

Bending stresses in beams with open thin-walled sections subjected to bending loads can be analyzed using the beam equations derived in previous sections with excellent accuracy if the beam span-to-depth ratio is large. In contrast, the transverse shear stresses τ_{xz} and τ_{xy} are very difficult to obtain. In fact, for a thin-walled section, τ_{xz} and τ_{xy}, in general, are not the most convenient stress components to consider. For instance, it is more advantageous to set up the s–n coordinate system for the thin-walled section shown in Fig. 5.1. The s-axis follows the center line of the wall, and the n-axis is perpendicular to the s-axis. Referring to the s–n coordinate system, the shear stresses can be represented by τ_{xn} and τ_{xs}, as shown in Fig. 5.1. Again, using the argument that the wall section is thin and that τ_{xn} vanishes on the boundary, we conclude that τ_{xn} must be small, and we may set $\tau_{xn} = 0$ over the entire wall section as an approximation. Thus, the only nonvanishing

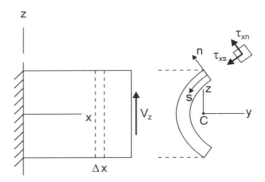

Figure 5.1   Thin-walled section symmetrical with respect to the y-axis.

shear stress component is τ_{xs} (or simply τ), and a great simplification is achieved.

5.1.1   Symmetric Thin-Walled Sections

If the cross-section is symmetrical about the y- or z-axis, then $I_{yz} = 0$. For one-way bending, say $V_z \neq 0$ and $V_y = 0$, beam equations (4.39)–(4.41) are to be used. Consider the beam section given by Fig. 5.1 which is symmetrical about the y-axis. We set up the s–n coordinate system as shown in Fig. 5.1. The shear stress τ_{xs} is assumed to vanish along the longitudinal edges.

Take a free body cut from the beam as shown in Fig. 5.2. The balance of forces in the x-direction of this free body requires that

$$\iint_{A_s} \Delta \sigma_{xx} \, dA = -q_s \, \Delta x$$

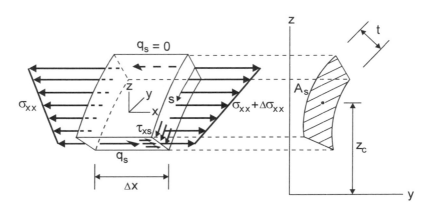

Figure 5.2   Free body cut from a beam.

where $q_s = t\tau_{xs}$ is the flexural shear flow on the thin-walled section and A_s is the cross-sectional area of the free body. Taking limit $\Delta x \to 0$, we have

$$\iint_{A_s} \frac{d\sigma_{xx}}{dx} \, dA = -q_s \tag{5.1}$$

Noting that

$$\sigma_{xx} = \frac{M_y z}{I_y} \quad \text{and} \quad \frac{dM_y}{dx} = V_z$$

we derive the following expression from (5.1):

$$q_s = -\frac{V_z}{I_y} \iint_{A_s} z \, dA$$

$$= -\frac{V_z Q}{I_y} \tag{5.2}$$

where

$$Q = \iint_{A_s} z \, dA = A_s z_c$$

is the first moment of area A_s, and z_c is the vertical distance from the centroid of A_s to the y-axis.

Comparing (5.2) with (4.47), we note that q_z and q_s differ by a sign. This is because the assumed positive directions (q_s is positive in the positive s-direction) of the two shear flows are opposite. The shear flow calculated according to (5.2) is the flexural shear flow because it is induced solely by the bending stress. The resultant force of the shear flow is equal to the applied shear force V_z.

Note that in the foregoing derivation of the flexural shear flow, the horizontal position of V_z is never specified. This is because bending moment M_y does not depend on the horizontal position of the shear force V_z. However, this does not imply that V_z can be applied at any arbitrary position if torsional shear stresses are to be avoided. This subject is discussed further in Section 5.2.

Example 5.1 The beam with the channel section shown in Fig. 5.3a is loaded with a constant shear force V_z ($V_y = 0$). The wall thickness is t. Since the section is symmetrical about the y-axis, the flexural shear flow can be calculated using (5.2).

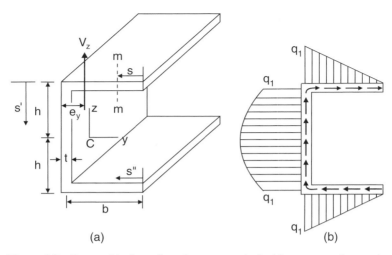

Figure 5.3 Beam with channel section symmetrical with respect to the y-axis.

Consider an arbitrary section m–m in the upper flange. For this section we have

$$A_s = ts, \qquad z_c = h$$

Thus,

$$q_s = -\frac{V_z t h s}{I_y}, \qquad 0 \le s \le b$$

This indicates that the shear flow is linearly distributed along the upper flange, and the negative sign means that its direction is opposite that of the contour s, as shown in Fig. 5.3b. The maximum value occurs at $s = b$, i.e.,

$$(q_s)_{\max} = -\frac{V_z t b h}{I_y} = -q_1$$

In a similar manner, the shear flow on the vertical web is obtained as

$$q_s = -\frac{V_z[tbh + ts'(h - s'/2)]}{I_y}$$

where s' is measured starting from the top end of the web. The second term in the numerator in the equation above is the first moment of the cross-sectional area of the vertical web up to the point s'. At the top of the web, $s' = 0$ and

$$q_s = -\frac{V_z t b h}{I_y} = -q_1$$

At the midsection of the web, $s' = h$, and

$$q_s = -\frac{V_z t h (b + h/2)}{I_y} = -q_1 - \frac{V_z t h^2}{2 I_y} = -q_2$$

At the bottom of the web, $s' = 2h$, and

$$q_s = -q_1$$

Again, the negative value of q_s indicates that the actual direction of q_s is opposite that of s'.

The shear flow on the lower flange can be calculated using the same counterclockwise shear flow contour. Alternatively, we may choose a new clockwise contour s'' as shown in Fig. 5.3a. We have

$$A_s = t s'', \qquad z_c = -h$$

and

$$q_s = \frac{V_z t h s''}{I_y}, \qquad 0 \le s'' \le b$$

The shear flow along the lower flange given by the preceding equation is positive, indicating that its direction is the same as contour s''.

It can be verified that the resultant force of the shear flow is equal to V_z.

Stringer-Web Sections   For stringer-web constructions such as the one with the cross-section shown in Fig. 5.4, stringers are used to take bending. Often the web can be assumed to be ineffective in bending, and its area

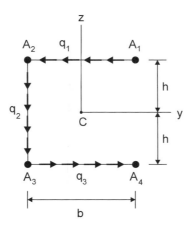

Figure 5.4   Stringer–web section symmetrical with respect to the y-axis.

neglected in the calculation of Q. As a result, the shear flow between two adjacent stringers becomes constant.

Example 5.2   Consider a four-stringer thin-walled channel beam with cross-section shown in Fig. 5.4. Assume that $A_3 = A_2$ and $A_4 = A_1$; then the section is symmetric about y-axis, and the shear flow equation (5.2) can be used. Thus, shear flow q_i produced by a vertical shear force V_z is obtained as

$$q_i = -\frac{V_z Q_i}{I_y}$$

where

$$Q_i = \sum_{k=1}^{i} z_k A_k$$

$$I_y = 2h^2 (A_1 + A_2)$$

and z_k is the vertical position of stringer A_k.
    For q_1, we have

$$Q_1 = A_1 h$$

and thus

$$q_1 = -\frac{V_z A_1 h}{2h^2(A_1 + A_2)}$$
$$= -\frac{V_z A_1}{2h(A_1 + A_2)}$$

Similarly, we have

$$q_2 = -\frac{V_z(A_1 + A_2)h}{2h^2(A_1 + A_2)} = -\frac{V_z}{2h}$$
$$q_3 = -\frac{V_z(A_1 h + A_2 h - A_2 h)}{2h^2(A_1 + A_2)}$$
$$= -\frac{V_z A_1}{2h(A_1 + A_2)} = q_1$$

Note that the actual direction of the shear flow is opposite that shown in Fig. 5.4.

It is easy to show that the shear flow can also be expressed as

$$q_{i+1} = q_i - \frac{V_z}{I_y} z_{i+1} A_{i+1}$$

where z_{i+1} is the z-coordinate of area A_{i+1}.

The preceding result indicates that

$$V_z = -2hq_2$$

and

$$\sum F_y = bq_3 - bq_1 = 0$$

that is, the vertical resultant of the shear flow must be equal to the applied transverse shear force V_z, and the horizontal resultant force must vanish as there is no horizontally applied force.

5.1.2 Unsymmetric Thin-Walled Sections

For unsymmetric thin-walled sections under bidirectional bending, the equilibrium equation (5.1) is still valid. However, the bending stress σ_{xx} must be calculated from (4.29). For convenience, we rewrite (4.29) in the form

$$\sigma_{xx} = (k_y M_z - k_{yz} M_y)y + (k_z M_y - k_{yz} M_z)z \tag{5.3}$$

where

$$k_y = \frac{I_y}{I_y I_z - I_{yz}^2}, \qquad k_z = \frac{I_z}{I_y I_z - I_{yz}^2}, \qquad k_{yz} = \frac{I_{yz}}{I_y I_z - I_{yz}^2} \tag{5.4}$$

Substituting (5.3) together with (4.34b) and (4.34d) into (5.1), we obtain

$$q_s = -(k_y V_y - k_{yz} V_z)Q_z - (k_z V_z - k_{yz} V_y)Q_y \tag{5.5}$$

where

$$Q_z = \iint\limits_{A_s} y \, dA, \qquad Q_y = \iint\limits_{A_s} z \, dA \tag{5.6}$$

are the first moments of the area A_s about the z and y axes, respectively.

Example 5.3   Consider the stringer–web beam shown in Fig. 5.5. The shear flow produced by combined vertical load V_z and horizontal load V_y can be solved by considering these two loads separately. Consider the applied load

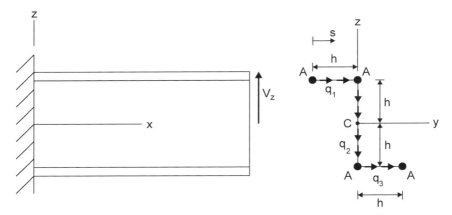

Figure 5.5   Stringer–web beam with an unsymmetrical section.

$V_z \neq 0$ and $V_y = 0$. The direction of the shear flow s is indicated in the figure. The positive direction of the shear flow is assumed to be the same as that of the contour s. The moments of inertia and product of inertia of the section are first calculated as

$$I_y = 4Ah^2, \qquad I_z = 2Ah^2, \qquad I_{yz} = -2Ah^2$$

Subsequently, we calculate k_y, k_z, and k_{yz} according to (5.4) with the result

$$k_y = \frac{1}{Ah^2}, \qquad k_z = \frac{1}{2Ah^2}, \qquad k_{yz} = -\frac{1}{2Ah^2}$$

For shear flow q_1, only one stringer is involved. We have

$$Q_y = Ah, \qquad Q_z = -Ah$$

and the shear flow is obtained from (5.5) as

$$q_1 = k_{yz} V_z Q_z - k_z V_z Q_y$$
$$= 0$$

For q_2, we have

$$Q_y = 2Ah, \qquad Q_z = -Ah$$

and

$$q_2 = \left(-\frac{1}{2Ah^2}\right)(-Ah)\,V_z - \frac{1}{2Ah^2}(2Ah)V_z$$
$$= -\frac{1}{2h}V_z$$

In the same manner, we obtain

$$q_3 = 0$$

Of course, in this case, constant shear flow q_2 can be obtained from the fact that the resultant transverse shear force must be equal to V_z, i.e.,

$$2hq_2 = -V_z$$

Again, the negative sign in the equation above indicates that the actual direction of the shear flow is opposite that assumed in Fig. 5.5.

5.1.3 Multiple Shear Flow Junctions

In multicell thin-walled sections, there are junctions where three or more shear flows meet. For example, consider the junction of three walls as shown in Fig. 5.6. We have shown that for shear flows produced by a torque,

$$q_3 = q_1 - q_2 \quad \text{or} \quad q_1 = q_2 + q_3 \tag{5.7}$$

This relation is valid for flexural shear flows produced by transverse forces for sections without concentrated areas. Relation (5.7) is obvious from the consideration of balance of force (in x-direction) for the free body of Fig. 5.6.

For sections consisting of stringers and thin sheets, the relation (5.7) is not valid. Consider the junction of three sheets and a stringer as shown in Fig. 5.7. The equilibrium equation in the x-direction is

$$\sum F_x = 0: \quad (\sigma_{xx} + \Delta\sigma_{xx})A - \sigma_{xx}A + q_1\,\Delta x - q_2\,\Delta x - q_3\,\Delta x = 0$$

Taking $\Delta x \to 0$, we have

$$q_1 = q_2 + q_3 - A\frac{d\sigma_{xx}}{dx} \tag{5.8}$$

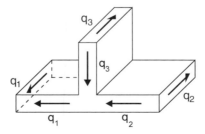

Figure 5.6 Junction of three walls.

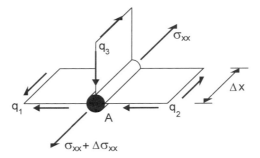

Figure 5.7   Junction of three sheets and a stringer.

For example, consider a symmetrical section subjected to transverse force V_z. We have

$$\sigma_{xx} = \frac{M_y z_c}{I_y}, \qquad \frac{dM_y}{dx} = V_z \tag{5.9}$$

where z_c is the vertical position of the stringer. Substitution of (5.9) into (5.8) yields

$$q_1 = q_2 + q_3 - \frac{V_z A z_c}{I_y} \tag{5.10}$$

which is different from (5.7).

5.1.4   Selection of Shear Flow Contour

In using (5.2) or (5.5) to calculate flexural shear flows, it is important to specify the shear flow contour s since it indicates the assumed direction of the shear flow. Other than the requirement that a contour must begin from a free edge, the choice can be arbitrary. For simple sections, such as those shown in Figs. 5.4 and 5.5, a single contour is sufficient, and we may select the contour to start from either the top free edge or the bottom free edge.

For some sections, it is convenient to set up different shear flow contours for different portions of the section. For example, consider the wide flange beam as shown in Fig. 5.8. Five contours are selected (see Fig. 5.8). Contours s_1, s_2, s_4, and s_5 can be used to calculate the shear flows in the flanges independently. Contour s_3 should be considered as the combined contour of s_1 and s_2. When calculating the shear flow in the vertical web (along s_3), the areas of both top flanges must be included. Using the relation $q_3 = q_1 + q_2$ at the junction, the shear flow along the vertical web can be calculated by adding the contribution of the additional area of the vertical web to the shear flows q_1 and q_2 in the top flanges at the junction.

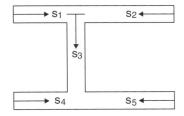

Figure 5.8   Five possible contours for shear flow.

5.2   SHEAR CENTER IN OPEN SECTIONS

In calculating the shear flow in an open section produced by shear forces, only the magnitude of the shear force is used. The position of the applied force in the y–z plane is not specified. However, the flexural shear flow resulting from the shear force has a definite resultant force location. This location is called the **shear center** of the cross-section. The shear center is sometimes called the center of twist. If a torque is applied about the shear center, the beam will twist without bending. Conversely, if the shear force is applied through the shear center, the beam will bend without twisting.

To illustrate the procedure of finding the location of shear center, let us consider a thin-walled bar with two heavy flanges as shown in Fig. 5.9. Assume that the curved web is ineffective in bending. Consequently, the shear flow is constant between the flanges, and is obtained as

$$q = \frac{V_z}{h} \tag{5.11}$$

Using (3.40) and (5.11), it can be easily verified that the resultant R of the shear flow is qh which is equal to the applied shear force V_z. In fact, shear flows calculated from the shear flow formulas (5.2) and (5.5) always give the applied shear forces.

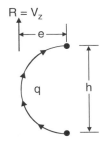

Figure 5.9   Thin-walled bar with two heavy flanges.

The location of the resultant force $R(= V_z)$ of the shear flow q is also the location of the shear center. The applied shear force V_z must be applied through this location in order to avoid additional torsional deformation.

Assuming that the shear center is at a distance e to the left of the top flange (see Fig. 5.9), and requiring that the moment produced by $R(= V_z)$ about the top flange be the same as by the shear flow, we have

$$V_z e = 2\overline{A} q = 2\overline{A} \frac{V_z}{h} \tag{5.12}$$

where

$$\overline{A} = \frac{1}{2} \pi \left(\frac{h}{2} \right)^2 = \frac{1}{8} \pi h^2 \tag{5.13}$$

is the area enclosed by the curved web and the straight line connecting the two flanges. From (5.12), we obtain

$$e = \frac{2\overline{A}}{h} = \frac{\pi h}{4} \tag{5.14}$$

The positive sign of e indicates that the location of the shear center is to the left of the flanges. Also note that the shear center location does not depend on the magnitude of the shear force.

In general, the location of a shear center is determined by its horizontal position and vertical position. The horizontal position is obtained from the loading condition $V_z \neq 0$ and $V_y = 0$, and the vertical position is determined using $V_y \neq 0$ and $V_z = 0$.

Example 5.4 The four-stringer thin-walled channel section of Fig. 5.10 is a special case of the section given in Fig. 5.4. To determine the horizontal position of its shear center, we consider the loading of $V_z \neq 0$ and $V_y = 0$. The resulting shear flow has been obtained in Example 5.2. We have

$$q_1 = \frac{V_z A_1}{2h(A_1 + A_2)}$$

$$q_2 = \frac{V_z}{2h}$$

$$q_3 = \frac{V_z A_1}{2h(A_1 + A_2)}$$

Note that the shear flow direction in Fig. 5.10 is the actual direction.

Let the resultant force of the shear flow be R_z which is obviously equal to V_z in magnitude and should pass through the shear center (see Fig. 5.10).

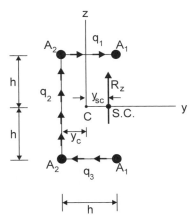

Figure 5.10   Shear-flow direction.

The resultant force R_z must also produce the same moment about the x-axis (or any other axis) as does the shear flow, i.e.,

$$R_z y_{sc} = -q_1(h)(h) - q_2(2h)(y_c) - q_3(h)(h) \qquad \text{(a)}$$

where

$$y_c = \frac{A_1 h}{A_1 + A_2}$$

is the horizontal distance between the centroid and the vertical web. Using $R_z = V_z$ and the expressions for q_1, q_2, and q_3, we obtain the distance y_{sc} from (a) as

$$y_{sc} = -\frac{2h A_1}{A_1 + A_2} \qquad \text{(b)}$$

The location $y = y_{sc}$ is the horizontal position of the **shear center** for this section for the shear force applied in the vertical (z) direction. The negative sign of y_{sc} in (b) indicates that the actual shear center is to the left of the centroid.

If the shear force V_z is applied through the shear center, this shear flow is the complete response of the structure. If the shear force V_z is not applied through the shear center as shown in Fig. 5.11a, then an additional torque load results. As shown in Fig. 5.11b, the shear force V_z can be translated to the shear center, resulting in a torque $V_z d$. In such cases, the shear stresses produced by the torque must be added to the flexural shear stresses. Since open thin-walled sections are generally weak in torsion, it is desirable to apply the shear force through the shear center.

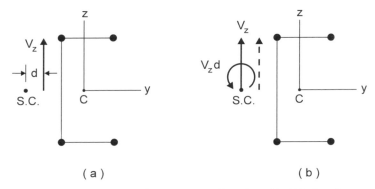

Figure 5.11 Location of shear force: (a) actual location; (b) shifted to shear center.

To find the vertical position z_{sc} of the shear center, we consider the section of Fig. 5.10 subjected to loading $V_y \neq 0$ and $V_z = 0$. For this section $k_{yz} = 0$, and from (5.5) we have

$$q_s = -k_y V_y Q_z \tag{c}$$

The location of the centroid is indicated in Fig. 5.10 with

$$y_c = \frac{A_1 h}{(A_1 + A_2)}$$

The shear flow is easily obtained from (c):

$$q_1 = -k_y V_y A_1(h - y_c) = -k_y V_y \frac{A_1 A_2 h}{A_1 + A_2} = -q_0$$

$$q_2 = q_1 - k_y V_y A_2(-y_c) = 0$$

$$q_3 = q_2 - k_y V_y A_2(-y_c) = k_y V_y \frac{A_1 A_2 h}{A_1 + A_2} = -q_1 = q_0$$

The shear flow, after adjusting the sign for direction, is shown in Fig. 5.12. The shear flow is seen to be symmetrical with respect to the y-axis. The resultant R_y thus coincides with the y-axis and, consequently, the vertical location of the shear center is $z_{sc} = 0$.

Simple Rule for Determining the Shear Center The following rule can be used to locate the shear center for sections possessing symmetries. If a section (including both stringers and thin webs) is symmetric about an axis, then the shear center lies on this axis. For example, the sections of Figs. 5.13a and 5.13b are symmetrical with respect to the y-axis and thus $z_{sc} = 0$.

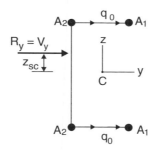

Figure 5.12 Shear flow for a horizontal shear force.

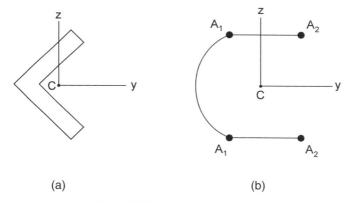

(a)   (b)

Figure 5.13 Symmetric sections.

This cannot be said about the section of Fig. 5.14. *Although it is considered a symmetrical section for bending when the thin webs are assumed ineffective in bending, the shear flow cannot be symmetric about the y-axis because of the unsymmetric webs.*

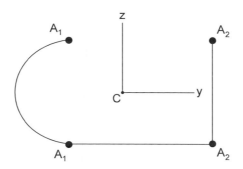

Figure 5.14 Unsymmetric section.

For sections such as the angle section in Fig. 5.13a, the T-section in Fig. 3.10a, and cruciform sections in which all walls meet at a single location, the shear center coincides with the intersection point. This conclusion is easily reached in view of the fact that all shear flows are straight and meet at one point.

Example 5.5   The beam with a channel section shown in Fig. 5.3 is symmetrical about the y-axis. Thus, $z_{sc} = 0$. To determine the horizontal position of the shear center, we consider the loading case $V_z \neq 0$, $V_y = 0$ for which the shear flow has already been obtained in Example 5.1 and is reproduced in Fig. 5.15b.

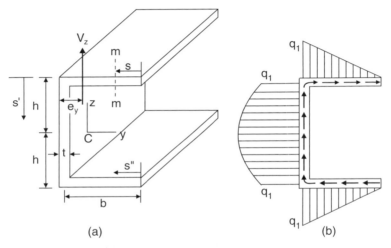

(a)   (b)

Figure 5.15   Reproduction of shear flow in Example 5.3.

Assume that V_z passes through the shear center which is assumed to be at e_y to the right of the vertical well. Then the moment produced by V_z and the shear flow about any axis (that is parallel to the x-axis) must be equal. By selecting the axis location at the lower left corner of the channel, the shear flows on the vertical wall and the lower flange produce no moment; and only the shear flow on the upper flange does. The shear flow on the upper flange can be written in the form

$$q = \frac{s}{b}q_1$$

The moment of this shear flow about the axis selected is clockwise and is given by

$$\int_0^b 2hq\,ds = \frac{hb^2q_1}{b} = hbq_1$$

For the assumed loading position (see Fig. 5.15), the moment produced by V_z about the same axis is $V_z e_y$, which is counterclockwise. Hence,

$$V_z e_y = -hbq_1$$

Noting that

$$q_1 = \frac{V_z tbh}{I_y}$$

we obtain

$$e_y = -\frac{tb^2h^2}{I_y}$$

The minus sign indicates that the actual location of the shear center is to the left of the vertical wall. Thus, a fitting sticking out from the vertical wall may be necessary to facilitate such loading.

5.3 CLOSED THIN-WALLED SECTIONS AND COMBINED FLEXURAL AND TORSIONAL SHEAR FLOW

Closed thin-walled sections are capable of taking both shear forces and torques. Shear flows can result from simultaneous applications of shear forces and torques. In the derivation of flexural shear flows in open sections, the flexural shear stress τ_{xs} (and thus, q_s) is zero at the free edges (see Fig. 5.16). For closed sections, such as shown in Fig. 5.17a, there are no free edges. We assume that at point O the value of the shear flow is q_0 (see Fig. 5.17b). Thus, the closed section can be regarded as an open section with a nonzero shear flow at point O. Starting contour s from this point (see Fig. 5.17b), we obtain the shear flow q_s as

$$q_s = q_s' + q_0 \tag{5.15}$$

where q_s' is the shear flow calculated assuming a free edge at point O. Hence, the actual shear flow can be considered as the superposition of $q_s'(s)$ and the unknown constant shear flow q_0 as depicted in Fig. 5.18. The flexural shear flow q' can be regarded as the shear flow produced by the shear force in the open section obtained by cutting the wall longitudinally at point O.

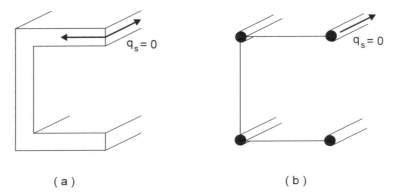

Figure 5.16 Flexural shear flows in open sections.

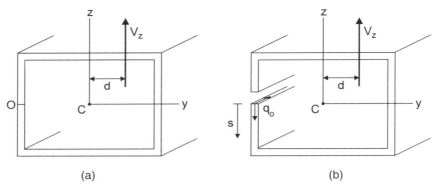

Figure 5.17 Flexural shear flows in a closed section: (a) closed section; (b) section with a fictitious cut.

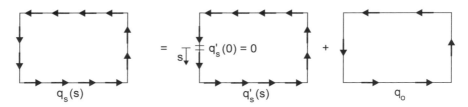

Figure 5.18 Superposition of shear flows.

The shear flow q_s can be viewed from the front section (positive x-face) or the back section (negative x-face). Viewed from the front section, the shear flow q_s should produce the resultant shear force equal to V_z. It should also generate the same moment as V_z about any axis parallel to the x-axis. Using the closed section and loading of Fig. 5.17 as an example, it is easy to show that the shear flow q_s calculated from (5.15) yields the resultant force V_z automatically. In fact, it is the shear flow q'_s that produces the shear force since the resultant force of the closed constant shear flow q_0 vanishes. The remaining condition is the momemt equation. Taking moment about the x-axis, we have

$$V_z \cdot d = 2\overline{A}q_0 + \text{moment produced by } q'_s \text{ about the } x\text{-axis} \qquad (5.16)$$

where $\overline{A}$ is the area enclosed by the shear flow.

Equation (5.16) ensures that the resulting shear flow must produce the same moment as the applied shear force V_z. This equation is used to determine q_0.

5.3.1 Shear Center

The shear flow given by (5.15) may contain flexural shear and torsional shear if V_z is applied at an arbitrary location. If the applied shear force V_z passes through the shear center, i.e., $d = y_{sc}$, then the resulting shear flow is pure flexural shear which should produce no twist angle, i.e.,

$$\theta = 0 = \frac{1}{2G\overline{A}} \oint \frac{q}{t} ds \qquad (5.17)$$

Equation (5.16) is used to determine q_0 in terms of y_{sc}. The location (y_{sc}) of the shear center, if not given, is subsequently obtained from solving (5.17) by replacing d with y_{sc}.

An equivalent problem to that of Fig. 5.17 can be obtained by translating the shear force V_z from $y = d$ to $y = y_{sc}$ (the shear center) and adding a torque $T = V_z(d - y_{sc})$ as shown in Fig. 5.19. The shear flow resulting from this torque must be added to the shear flow produced by the shear force that passes through the shear center.

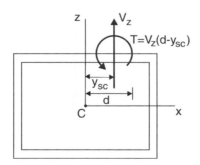

Figure 5.19   Added torque due to shifting of the shear force.

Example 5.6   A four-stringer box beam is loaded as shown in Fig. 5.20. Assume the thin sheets to be ineffective in bending. The centroid is easily identified and is shown in the figure.

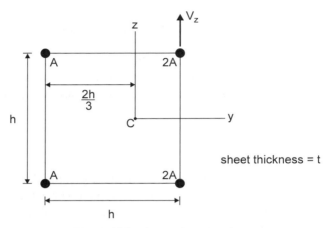

Figure 5.20   Four-stringer box beam.

As far as bending is concerned, this cross-section is symmetrical with respect to the y-axis. Thus, $I_{yz} = 0$. The other properties of the cross-section are given by

$$I_y = \tfrac{3}{2}Ah^2$$

$$I_z = \tfrac{4}{3}Ah^2$$

We first calculate the shear flow q' by assuming a cut (see Fig. 5.21) in the wall between stringers 1 and 2, i.e., $q'_{12} = 0$. The shear flows on other walls are calculated according to (5.2) for symmetrical sections. We obtain

$$q'_{23} = -\frac{V_z A(h/2)}{\tfrac{3}{2}Ah^2} = -\frac{V_z}{3h}$$

$$q'_{34} = 0$$

$$q'_{41} = -\frac{V_z(2A)(-h/2)}{\tfrac{3}{2}Ah^2} = \frac{2V_z}{3h}$$

The resulting moment of the total shear flow $q = q' + q_0$ must be equal to the moment produced by V_z. Taking moment about stringer 1 and using

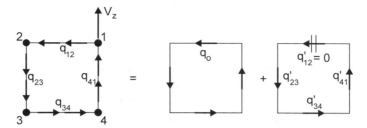

Figure 5.21   Superposition of shear flows.

(5.16), we obtain

$$V_z \cdot 0 = q'_{23}hh + 2\overline{A}q_0$$
$$= q'_{23}h^2 + 2h^2 q_0 \tag{a}$$

Thus,

$$q_0 = -\frac{1}{2}q'_{23} = \frac{V_z}{6h}$$

The total shear flows are

$$q_{12} = q'_{12} + q_0 = \frac{V_z}{6h}$$

$$q_{23} = q'_{23} + q_0 = -\frac{V_z}{6h}$$

$$q_{34} = q'_{34} + q_0 = \frac{V_z}{6h}$$

$$q_{41} = q'_{41} + q_0 = \frac{5V_z}{6h}$$

To determine the horizontal location of the shear center, we assume that the shear force V_z is applied through the shear center (assumed to be at distance e from stringer 1 as shown in Fig. 5.22). Then the moment equation (a) is replaced by

$$V_z e = q'_{23}h^2 + 2h^2 q_0$$

Thus,

$$q_0 = \frac{V_z e}{2h^2} - \frac{q'_{23}}{2} = \frac{V_z}{6h^2}(h + 3e) \tag{b}$$

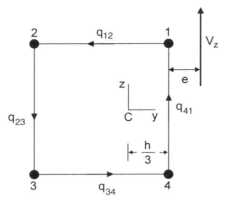

Figure 5.22   Assuming shear force passes through the shear center.

and

$$q_{12} = q_0 = \frac{V_z}{6h^2}(h + 3e)$$

$$q_{23} = \frac{V_z}{6h^2}(-h + 3e)$$

$$q_{34} = \frac{V_z}{6h^2}(h + 3e)$$

$$q_{41} = \frac{V_z}{6h^2}(5h + 3e)$$

Since V_z passes through the shear center, the twist angle is equal to zero. Using (5.17), we have

$$\theta = \frac{1}{2G\overline{A}}\left(q_{12}\frac{h}{t} + q_{23}\frac{h}{t} + q_{34}\frac{h}{t} + q_{41}\frac{h}{t}\right) = 0$$

This equation reduces to

$$q_{12} + q_{23} + q_{34} + q_{41} = 0$$

Solving the above equation for e, we obtain

$$e = -\tfrac{1}{2}h$$

The negative sign indicates that the shear center is located to the left of the vertical wall between stringers 1 and 4. It is obvious from Fig. 5.22 that

$$y_{sc} = \frac{h}{3} + e = -\frac{h}{6}$$

The vertical location z_{sc} of the shear center can be determined in a similar manner by applying a horizontal shear force V_y. The result is $z_{sc} = 0$, i.e., the shear center lies on the axis of symmetry of the cross-section.

5.3.2 Statically Determinate Shear Flow

At any cross-section of a thin-walled beam, the shear flow must result in the same resultant forces and moment as the applied ones, i.e.,

$$\sum F_y = V_y \tag{5.18a}$$

$$\sum F_z = V_z \tag{5.18b}$$

$$\sum M = V_y e_z + V_z e_y \tag{5.18c}$$

where e_z and e_y are the distances of V_y and V_z from the axis about which the moments are taken. For some sections, the shear flow can be determined from these equations alone. This type of shear flow is statically determinate. In this case, the sectional properties (I_y, I_z, and I_{yz}) are not involved.

Example 5.7   Consider a three-stringer single-cell section loaded as shown in Fig. 5.23. The three equations on the equivalent resultants are given by

$$\sum F_y = 0: \qquad hq_1 - hq_2 = 0 \tag{a}$$

$$\sum F_z = V_z: \qquad 2hq_3 - hq_1 - hq_2 = V_z \tag{b}$$

$$\sum M_1 = V_z e_y: \qquad \pi h^2 q_1 = V_z e_y \quad [q_1 = q_2 \text{ from (a) is used}] \tag{c}$$

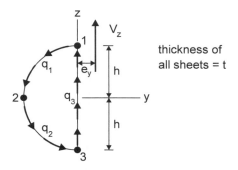

Figure 5.23   Three-stringer single-cell section.

Solving the equations above, we obtain

$$q_1 = q_2 = \frac{V_z e_y}{\pi h^2}$$

$$q_3 = \frac{(\pi h + 2e_y) V_z}{2\pi h^2}$$

If V_z passes through the shear center, then the twist angle $\theta = 0$, i.e.,

$$\frac{q_1(\pi h/2)}{t} + \frac{q_2(\pi h/2)}{t} + \frac{q_3(2h)}{t} = 0$$

This leads to

$$e_y = -\frac{\pi h}{2 + \pi}$$

The negative sign indicates that the shear center is located to the left of the vertical web.

Example 5.8   The cross-section of a three-stringer thin-walled beam and the applied loads are shown in Fig. 5.24a. This problem can be solved by first converting the loads into the resultant shear forces as shown in Fig. 5.24b and then using the method of cutting the closed cell into an open cell to find q'_S.

Alternatively, by recognizing that the shear flow is statically determinate, we can use (5.18) to determine the shear flow. Assuming constant shear flows in the thin-wall segments as shown in Fig. 5.25, we have

$$3P_h = bq_{23} - bq_{31} \tag{a}$$

$$3P_v = -hq_{12} + hq_{31} \tag{b}$$

$$2P_h h + P_v b = 2\overline{A}_1 q_{12} + 2\overline{A}_2 q_{23} \tag{c}$$

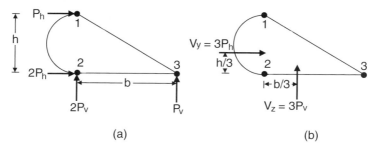

(a)   (b)

Figure 5.24   Resultant shear forces.

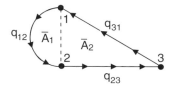

Figure 5.25   Assumed shear flows.

The left-hand side and right-hand side of (c) represent moments of the applied forces and the shear flow about stringer 1, respectively. In deriving the equations above, the relation given by (3.50) has been used. Also, the areas $\overline{A}_1$ and $\overline{A}_2$ are

$$\overline{A}_1 = \frac{1}{2}\pi \left(\frac{h}{2}\right)^2 = \frac{1}{8}\pi h^2$$

$$\overline{A}_2 = \frac{1}{2}hb$$

The shear flows q_{12}, q_{23}, and q_{31} are obtained by solving the three equations (a), (b), and (c).

5.4   CLOSED MULTICELL SECTIONS

As discussed in Section 5.3, the shear flow in a single-cell beam can be analyzed by making a fictitious cut so that it can be treated as an open section with an existing constant shear flow q_0. The shear flow q' is unambiguously obtained from (5.2) or (5.5) for the "open section" subjected to the applied shear forces. The unknown shear flow q_0 is determined from the requirement that the moment produced by the total shear flow $q' + q_0$ must be equal to the moment produced by the applied shear forces.

The aforementioned procedure can be employed for the analysis of shear flows in beams with multicell thin-walled cross-sections. For instance, consider an n-cell section. Make a "cut" in the wall in each cell to make the entire section "open." For each cell, a constant shear flow q_i ($i = 1, 2, \ldots, n$) must be added to the shear flow q' calculated for the open section. It requires n equations to solve for the n unknowns q_i. These n equations are provided by the $n - 1$ compatibility equations,

$$\theta_1 = \theta_2 = \cdots = \theta_n \tag{5.19}$$

where θ_i is the twist angle per unit length of the ith cell. An additional equation is provided by equating the moment of the applied shear forces to the total resultant moment of all the shear flows in the cells.

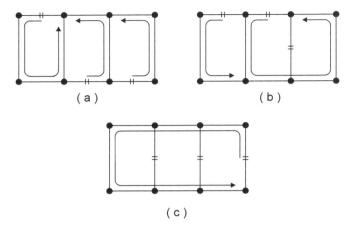

Figure 5.26   A few possible ways to cut a three-cell box beam section.

In making the cuts, no part of the section should be completely cut off. In setting up the shear flow contours for the resulting open section, it is more convenient to begin each contour from the cut location where $q' = 0$. Moreover, each wall can be covered by only a single contour. Figure 5.26 shows a few possible ways to cut a three-cell box beam section. Apparently, the cut depicted in Fig. 5.26c is the most convenient because a single contour is sufficient.

Example 5.9   The two-cell box beam section shown in Fig. 5.27 is symmetrical about the y-axis.

Assume that the sheets are ineffective in bending. The pertinent cross-sectional property is

$$I_y = 2(1 + 2 + 3)(20)^2 = 4800 \text{ cm}^4$$

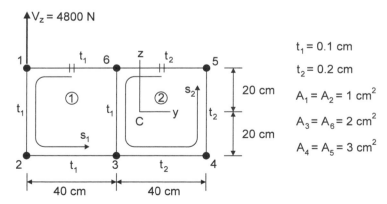

Figure 5.27   Assumed cuts and shear flow contours.

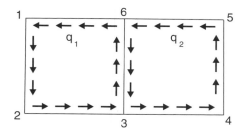

Figure 5.28   Constant shear flows to be added to the total shear flow.

Cut the sheets between stringers 1 and 6 and between 5 and 6, and set up the contours s_1 and s_2 as shown in Fig. 5.27. Each contour must start from the cut (the free edge). The positive shear flow direction is assumed to be in the contour direction. In the following, q_{ij} is used to denote the shear flow between stringers i and j.

Cell 1:

$$q'_{61} = 0$$

$$q'_{12} = -\frac{4800}{4800}(1)(20) = -20 \text{ N/cm} \tag{a}$$

$$q'_{23} = q'_{12} - \frac{4800}{4800}(1)(-20) = 0$$

Cell 2:

$$q'_{56} = 0$$

$$q'_{63} = -\frac{4800}{4800}(2)(20) = -40 \text{ N/cm}$$

$$q'_{34} = q'_{23} + q'_{63} - \frac{4800}{4800}(2)(-20) = 0 \tag{b}$$

$$q'_{45} = q'_{34} - \frac{4800}{4800}(3)(-20) = 60 \text{ N/cm}$$

The shear flows are completed by adding the constant shear flows q_1 and q_2 (see Fig. 5.28) in the individual cells, respectively. The equations needed for determining q_1 and q_2 are obtained from the moment equation and the compatibility equation.

Moment Equation   The in-plane moment produced by V_z about any axis must be equal to the in-plane moment about the same axis resulting from the shear flows. Taking the moment about stringer 1, we have

$$V_z \cdot 0 = 2\overline{A}_1 q_1 + 2\overline{A}_2 q_2 + q'_{23}(40)(40) + q'_{34}(40)(40)$$
$$+ q'_{45}(40)(80) - q'_{63}(40)(40) \tag{c}$$

where

$$\overline{A}_1 = \overline{A}_2 = (40)(40) = 1600 \text{ cm}^2$$

Substituting the numerical values of q'_{jj} given by (a) and (b) into (c), we obtain

$$q_1 + q_2 = -80 \text{ N/cm} \tag{d}$$

Compatibility Equation   The compatibility condition requires that the twist angle of cell 1 must be equal to that of cell 2. Using (3.56), we have

$$\frac{1}{2\overline{A}_1 G} \left(\frac{40q_{61}}{t_1} + \frac{40q_{12}}{t_1} + \frac{40q_{23}}{t_1} - \frac{40q_{63}}{t_1} \right)$$
$$= \frac{1}{2\overline{A}_2 G} \left(\frac{40q_{56}}{t_2} + \frac{40q_{63}}{t_1} + \frac{40q_{34}}{t_2} + \frac{40q_{45}}{t_2} \right) \tag{e}$$

where $\overline{A}_1$ and $\overline{A}_2$ are the areas enclosed by the centerlines of the thin walls in cells 1 and 2, respectively, and

$$q_{61} = q_1$$
$$q_{12} = q'_{12} + q_1 = q_1 - 20$$
$$q_{23} = q'_{23} + q_1 = q_1$$
$$q_{63} = q'_{63} - q_1 + q_2 = -40 - q_1 + q_2$$
$$q_{34} = q'_{34} + q_2 = q_2$$
$$q_{45} = q'_{45} + q_2 = 60 + q_2$$
$$q_{56} = q'_{56} + q_2 = q_2$$

Equation (e) is simplified to

$$10q_1 - 7q_2 = -60 \tag{f}$$

Solving (d) and (f), we obtain

$$q_1 = -36.47 \text{ N/cm}, \qquad q_2 = -43.53 \text{ N/cm}$$

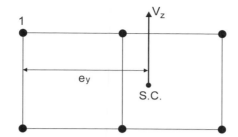

Figure 5.29   Applied force passing through a shear center.

Shear Center   To find the shear center, we assume that the applied force passes through the shear center as shown in Fig. 5.29. The resultant torque of the shear flow and the torque produced by V_z must be equal. Taking the moment about stringer 1, we have

$$V_z e_y = \sum M_1$$

where $\sum M_1$ is the same as the right-hand side of (c). Explicitly, the equation above is given by

$$V_z e_y = 3200(q_1 + q_2) + 256,000 \tag{g}$$

By the definition of shear center, we require that

$$\theta_1 = 0 = q_{61} + q_{12} + q_{23} - q_{63} \tag{h}$$

$$\theta_2 = 0 = \frac{q_{56}}{t_2} + \frac{q_{63}}{t_1} + \frac{q_{34}}{t_2} + \frac{q_{45}}{t_2} \tag{i}$$

These three equations are sufficient to solve for q_1, q_2, and the shear center location e_y. The solutions are

$$q_1 = -4.44 \text{ N/cm}$$

$$q_2 = 2.22 \text{ N/cm}$$

$$e_y = 51.85 \text{ cm}$$

PROBLEMS

5.1 Find the flexural shear flow produced by the transverse shear force V_z = 1000 N in the beam with the thin-walled section given by Fig. 5.30.

5.2 Find the shear flow of the wide-flange beam (Fig. 5.31) subjected to $V_z = 1000$ N.

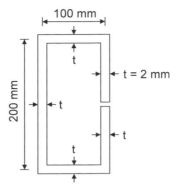

Figure 5.30   Thin-walled section with a side cut.

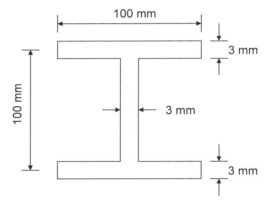

Figure 5.31   Section of an I-beam.

5.3 Find the shear center y_{sc} for the sections of Figs. 5.30 and 5.32. For the four-stringer section (Fig. 5.32), assume that the thin sheets are ineffective in bending.

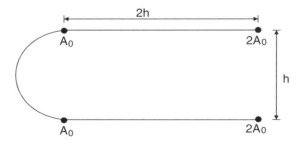

Figure 5.32   Open four-stringer section.

5.4 Find the flexural shear flow in the section of Fig. 5.32 for $V_z = 5000$ N.

5.5 Find the shear flow for the three-stringer section shown in Fig. 5.33 for $V_z = 5000$ N and $V_y = 0$. Given shear modulus $G = 27$ GPa, find the twist angle per unit length. Also determine the shear center. Is the shear flow statically determinate?

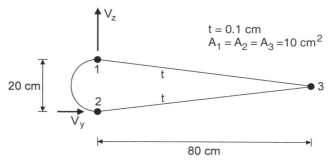

Figure 5.33   Single-cell closed section.

5.6 Do Problem 5.5 for $V_z = 5000$ N and $V_y = 10,000$ N.

5.7 Find the shear flow on the four-stringer section (Fig. 5.34) subjected to $V_z = 5000$ N. Assume that the thin sheets are ineffective in bending.

5.8 Find the shear center (y_{sc}, z_{sc}) for the open section in Fig. 5.34.

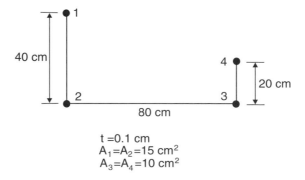

Figure 5.34   Unsymmetric open section.

5.9 Find the shear center of the Z-section given by Fig. 4.8.

5.10 Find the shear flow in the two-cell section loaded as shown in Fig. 5.35 for $V_z = 5000$ N. Given $G = 27$ GPa, find the twist angle θ.

5.11 Find the shear flow of the structure with the cross-section given in Fig. 5.35 if the vertical force V_z is applied at 20 cm to the right of the stringers. Also find the corresponding angle of twist θ.

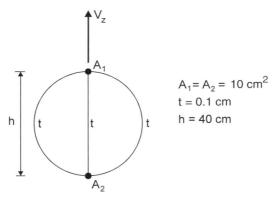

Figure 5.35   Two-cell closed section.

5.12 Solve Example 5.9 by assuming cuts on the webs between stringers 1 and 2 and stringers 6 and 3.

5.13 A thin-walled box beam is obtained by welding the cut of the section shown in Fig. 5.30. Find the shear flow produced by a vertical shear force $V_z = 1000$ N applied at 100 mm to the right of the vertical wall that contains the original cut.

5.14 Show that the shear center for the section of Fig. 5.36 is at a distance

$$e = \frac{a(a + \alpha b)}{(a + b)(1 + \alpha)}$$

to the left of stringer 1.

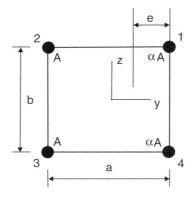

Figure 5.36   Four-stringer thin-walled section.

5.15 Find the shear flow in the two-cell thin-walled section for $V_z = 5000$ N shown in Fig. 5.37. Also determine the shear center. Assume thin sheets to be ineffective in bending.

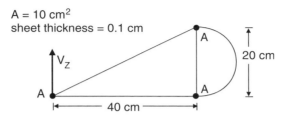

Figure 5.37 Two-cell closed section.

5.16 Find the shear flow in the five-stringer thin-walled section produced by the loads shown in Fig. 5.38.

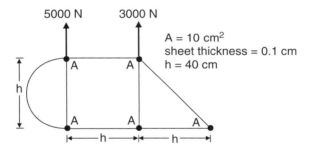

Figure 5.38 Three-cell closed section.

6

FAILURE CRITERIA FOR ISOTROPIC MATERIALS

The modes of failure of a structure can be put in two general categories, structural failure and material failure. The former is characterized by the loss of ability to perform the intended structural function. Examples of structural failure include elastic buckling and flutter. In general, a structural failure may be attributed to excessive deflections of the structure and may not necessarily involve breakage of the structure. On the other hand, material failure usually involves excessive permanent deformation or fracture of the material. In this chapter we address the latter category of failure. For convenience, a state of *plane stress* is assumed. In addition, only isotropic materials are considered.

6.1 STRENGTH CRITERIA FOR BRITTLE MATERIALS

In general, brittle materials exhibit linear stress-strain curves and have small strains to failure. Their uniaxial strengths can be determined by simple tension and compression tests. However, to predict failure in the material under a state of combined stresses, strength (or failure) criteria are needed in conjunction with these uniaxial strength data.

Many stress criteria for brittle materials have been proposed based on the phenomenological approach, which is basically an educated curve-fitting approach. Presented next are two criteria often employed to predict failure in brittle materials.

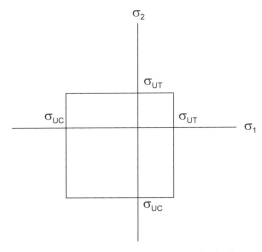

Figure 6.1 Failure envelope for the maximum principal stress criterion.

6.1.1 Maximum Principal Stress Criterion

Suppose that from simple tension and compression tests, one obtains ultimate strengths σ_{UT} and σ_{UC} (σ_{UC} is negative), respectively. Now consider a solid in a state of plane stress, and let σ_1 and σ_2 ($\sigma_3 = 0$) be the principal stresses in the plane. The **maximum principal stress criterion** states that failure would occur if

$$\sigma_1 \geq \sigma_{UT} \quad \text{for} \quad \sigma_1 > 0$$
$$\sigma_1 \leq \sigma_{UC} \quad \text{for} \quad \sigma_1 < 0 \tag{6.1a}$$

or

$$\sigma_2 \geq \sigma_{UT} \quad \text{for} \quad \sigma_2 > 0$$
$$\sigma_2 \leq \sigma_{UC} \quad \text{for} \quad \sigma_2 < 0 \tag{6.1b}$$

This failure criterion basically states that if any of the principal stresses reaches the ultimate strength, failure in the material would occur.

This failure criterion is presented graphically in Fig. 6.1. The line forming the square box in the σ_1–σ_2 plane is called the **failure envelope**. Any stress state within the envelope would not produce failure.

6.1.2 Coulomb–Mohr Criterion

The maximum normal stress criterion does not allow interaction of σ_1 and σ_2. However, many experiments have suggested that the presence of σ_2 could

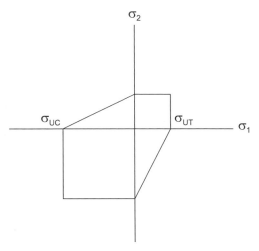

Figure 6.2   Coulomb–Mohr failure envelope.

reduce the ultimate value of σ_1. The **Coulomb–Mohr criterion** is among many criteria that attempt to account for the effect of stress interaction.

The Coulomb–Mohr failure envelope for a state of plane stress ($\sigma_3 = 0$) is shown in Fig. 6.2. The envelope is given by the following equations.

First quadrant ($\sigma_1 > 0,\ \sigma_2 > 0$) : $\sigma_1 = \sigma_{UT},\ \sigma_2 = \sigma_{UT}$ (6.2a)

Third quadrant ($\sigma_1 < 0,\ \sigma_2 < 0$) : $\sigma_2 = \sigma_{UC},\ \sigma_2 = \sigma_{UC}$ (6.2b)

Second quadrant ($\sigma_1 < 0,\ \sigma_2 > 0$) : $\dfrac{\sigma_2}{\sigma_{UT}} + \dfrac{\sigma_1}{\sigma_{UC}} = 1$ (6.2c)

Fourth quadrant ($\sigma_1 > 0,\ \sigma_2 < 0$) : $\dfrac{\sigma_1}{\sigma_{UT}} + \dfrac{\sigma_2}{\sigma_{UC}} = 1$ (6.2d)

For a pure shear produced by $\sigma_2 = -\sigma_1$ (assume that $\sigma_1 > 0$), the maximum shear stress is $\tau_U = \frac{1}{2}|\sigma_1 - \sigma_2| = \sigma_1$. From (6.2d) we have

$$\tau_U = \sigma_1 = \frac{\sigma_{UT}\sigma_{UC}}{\sigma_{UC} - \sigma_{UT}}$$

which is the shear stress at failure under pure shear loading.

Example 6.1   A thin-walled tube is made of a brittle material having $\sigma_{UT} = 200$ MPa and $\sigma_{UC} = -500$ MPa. The mean radius is $a = 0.2$ m and the wall thickness $t = 0.004$ m. Of interest is to find the maximum torque T that the tube can carry.

Under a torque, the state of the stress in the tube is pure shear. The shear stress τ can be obtained from the shear flow q which is related to the torque.

We have

$$T = 2\bar{A}q = 2\bar{A}t\tau = 2\pi a^2 t\tau$$
$$= 2\pi(0.2)^2 \times (0.004) \times \tau = 0.001\tau \tag{a}$$

It is easy to show that for pure shear, the principal stresses are $\sigma_1 = -\sigma_2 = \tau$. Without loss of generality, we assume that $\tau > 0$.

According to the maximum principal stress criterion, there are two possible failure loads, i.e.,

$$\tau = \sigma_1 = \sigma_{UT} = 200 \text{ MPa}$$

and

$$\tau = -\sigma_2 = -\sigma_{UC} = 500 \text{ MPa}$$

It is obvious that

$$\tau_{\max} = 200 \text{ MPa}$$

Thus, the maximum torque is, from (a),

$$T_{\max} = 0.001\tau_{\max} = 200 \text{ kN·m} \tag{b}$$

If the Coulomb–Mohr criterion is used, then the maximum shear stress is obtained from (6.2d). We have

$$\tau_{\max} = \frac{\sigma_{UT}\sigma_{UC}}{\sigma_{UC} - \sigma_{UT}} = 143 \text{ MPa}$$

and the corresponding maximum torque

$$T_{\max} = 143 \text{ kN·m} \tag{c}$$

The significant discrepancy between the predictions of these two strength criteria in this example serves to emphasize that the applicability of these strength criteria may vary from material to material and that great caution must be exercised in using these criteria.

6.2   YIELD CRITERIA FOR DUCTILE MATERIALS

Many materials exhibit substantial strain before fracture. When stressed beyond a level called **yield stress**, the material may exhibit inelastic behavior as shown in Fig. 6.3. Part of the strain produced beyond the yield stress σ_Y

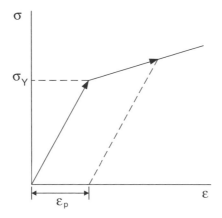

Figure 6.3 Inelastic behavior.

will remain even after stress is removed. This permanent strain ε_p is called plastic strain. Yielding can be considered a form of failure if no permanent deformation is allowed in the structure.

For uniaxial loading, the yield stress can be determined from the uniaxial stress-strain curve, i.e., yielding occurs if $\sigma \geq \sigma_Y$. For a state of more complex stresses, yield criteria are needed to determine whether permanent strains have been produced.

6.2.1 Maximum Shear Stress Criterion (Tresca Yield Criterion) in Plane Stress

At the atomic scale, plastic deformation is associated with sliding of adjacent layers of atoms. This slip action is referred to as *dislocation* which is produced by shearing of the solid. The **Tresca yield criterion** is proposed based on the assumption that if the shear stress exceeds the critical value, then dislocation, and thus yielding, would occur.

Consider a state of plane stress parallel to the $x-y$ plane. Denote σ_1 and σ_2 as the principal stresses in the $x-y$ plane, and $\sigma_3 = \sigma_{zz} = 0$. From Chapter 2, we note that the three local maximum shear stresses are given by

$$\tfrac{1}{2}|\sigma_1 - \sigma_2|, \qquad \tfrac{1}{2}|\sigma_1 - \sigma_3|, \qquad \tfrac{1}{2}|\sigma_2 - \sigma_3|$$

Let the critical value of shear stress be τ_Y. Then, noting that $\sigma_3 = 0$ from plane stress, the yield criterion can be expressed as

$$|\sigma_1 - \sigma_2| \geq 2\tau_Y \qquad (6.3a)$$

$$|\sigma_1| \geq 2\tau_Y \qquad (6.3b)$$

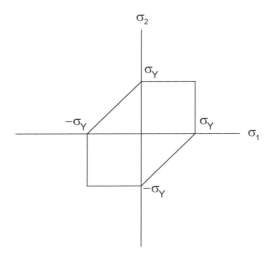

Figure 6.4   Yield surface for the Tresca yield criterion.

$$|\sigma_2| \geq 2\tau_Y \tag{6.3c}$$

Yielding occurs if any of the inequalities above is satisfied.

For simple tension, $\sigma_1 \neq 0$, $\sigma_2 = 0$, we know that yielding occurs at $\sigma_1 = \sigma_Y$. Substituting $\sigma_1 = \sigma_Y$ into (6.3a) or (6.3b), we obtain

$$\sigma_Y = 2\tau_Y$$

Thus, the Tresca yield criterion can also be written as

$$|\sigma_1 - \sigma_2| \geq \sigma_Y \tag{6.4a}$$

$$|\sigma_1| \geq \sigma_Y \tag{6.4b}$$

$$|\sigma_2| \geq \sigma_Y \tag{6.4c}$$

The yield surface (envelope) shown in Fig. 6.4 is constructed by the minimum values of the left-hand side of (6.4). Stresses inside the envelope produce no plastic strains.

6.2.2   Maximum Distortion Energy Criterion (von Mises Yield Criterion)

In an isotropic material, deformation can be separated into two parts, i.e., dilatation (or volume change) and distortion (or shape change). Plastic deformation resulting from dislocation accompanies distortion but not dilatation.

From Problem 2.10, we note that dilatation is given by

$$\varepsilon_0 = \varepsilon_{xx} + \varepsilon_{yy} + \varepsilon_{zz} = \frac{\Delta V}{V} \tag{6.5}$$

With the aid of the stress-strain relations given by (2.101), (6.5) can be expressed in terms of stress components as

$$\varepsilon_0 = \frac{3(1 - 2v)}{E}\sigma_0 \tag{6.6}$$

where

$$\sigma_0 = \tfrac{1}{3}(\sigma_{xx} + \sigma_{yy} + \sigma_{zz}) \tag{6.7}$$

is the average stress. Equation (6.6) can be written in terms of the **bulk modulus** K as

$$\sigma_0 = K\varepsilon_0 \tag{6.8}$$

where

$$K = \frac{E}{3(1 - 2v)}$$

For a state of plane stress, we have

$$\sigma_{zz} = 0, \qquad \varepsilon_{zz} = -v(\varepsilon_{xx} + \varepsilon_{yy}) \tag{6.9}$$

Thus, (6.5) and (6.7) reduce to

$$\varepsilon_0 = (1 - v)(\varepsilon_{xx} + \varepsilon_{yy})$$

$$\sigma_0 = \tfrac{1}{3}(\sigma_{xx} + \sigma_{yy})$$

The strain energy density (energy per unit volume) associated with the volume dilatation is given by

$$W_v = \frac{1}{2}\sigma_0\varepsilon_0$$

$$= \frac{1}{2K}\sigma_0^2$$

$$= \frac{1}{18K}(\sigma_{xx} + \sigma_{yy})^2 \tag{6.10}$$

The total strain energy density for plane stress is

$$W = \tfrac{1}{2}(\sigma_{xx}\varepsilon_{xx} + \sigma_{yy}\varepsilon_{yy} + \tau_{xy}\gamma_{xy}) \tag{6.11a}$$

Using the stress-strain relations (2.104), we have

$$W = \frac{1}{2E}\left(\sigma_{xx}^2 + \sigma_{yy}^2 - 2\nu\sigma_{xx}\sigma_{yy}\right) + \frac{1}{2G}\tau_{xy}^2 \qquad (6.11b)$$

The strain energy associated with distortional deformation is obtained as

$$W_d = W - W_v \qquad (6.12)$$

After some manipulations, (6.12) can be expressed explicitly in the form

$$W_d = \frac{1}{2G}J_2 \qquad (6.13)$$

where

$$J_2 = \tfrac{1}{6}[(\sigma_{xx} - \sigma_{yy})^2 + \sigma_{xx}^2 + \sigma_{yy}^2 + 6\tau_{xy}^2] \qquad (6.14)$$

If x and y axes are chosen parallel to the principal directions of stress, then $\tau_{xy} = 0$ and J_2 can be expressed in terms of the principal stresses as

$$J_2 = \tfrac{1}{6}[(\sigma_1 - \sigma_2)^2 + \sigma_1^2 + \sigma_2^2]$$
$$= \tfrac{1}{3}(\sigma_1^2 - \sigma_1\sigma_2 + \sigma_2^2) \qquad (6.15)$$

The **maximum distortion energy criterion** states that yielding begins if the distortion energy reaches a critical value W_0, i.e.,

$$W_d = \frac{1}{2G}J_2 = W_0 \qquad (6.16)$$

This critical value W_0 can be determined from substituting the known yielding condition in simple tension, $\sigma_{xx} = \sigma_Y$, into (6.16). We have

$$\frac{1}{2G} \times \frac{1}{3}\sigma_Y^2 = W_0$$

Since W_0 is a material constant, it can be used in the yield criterion for more general states of stress. Thus, yield criterion (6.16) becomes

$$J_2 = \tfrac{1}{3}\sigma_Y^2 \qquad (6.17)$$

This is also known as the **von Mises yield criterion** for isotropic materials. The yield surface is represented graphically in Fig. 6.5.

The von Mises yield criterion for isotropic solids in a state of plane strain can be derived in a similar manner. It can be expressed in the form of (6.17)

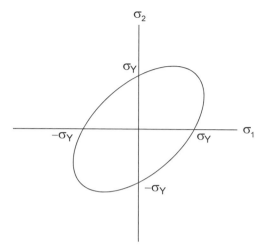

Figure 6.5   von Mises yield criterion.

with

$$J_2 = \tfrac{1}{3}\left[(1 - \nu + \nu^2)(\sigma_1 - \sigma_2)^2 + (1 - 2\nu)^2\sigma_1\sigma_2\right] \qquad (6.18)$$

Since the Poisson ratios for most isotropic solids are within the bounds of 0 and 0.5, it is not difficult to see from comparing (6.17) and (6.18) that the plane stress J_2 is larger than the plane strain J_2 if both σ_1 and σ_2 are positive (tensile). This implies that it is easier to produce yielding in a solid under plane stress than under plane strain.

Example 6.2   A thin-walled hollow cylinder is subjected to an axial force N, a torque T and an internal pressure p_0, as shown in Fig. 6.6. It is assumed

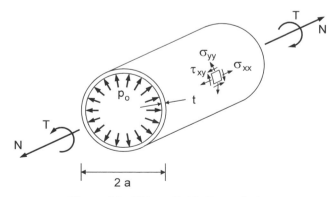

Figure 6.6   Thin-walled hollow cylinder.

that the yield stress is $\sigma_Y = 280$ MPa, the radius is $a = 1$ m, and wall thickness is $t = 5 \times 10^{-3}$ m.

The stresses produced by these loads are uniform over the entire cylinder. We use a local x–y coordinate system as shown in Fig. 6.6 to describe the stress field. The axial stress σ_{xx} is obtained by dividing the axial force N with the cross-sectional area of the thin wall $2\pi a t$:

$$\text{Axial stress:} \qquad \sigma_{xx} = \frac{N}{2\pi a t}$$

The shear stress is produced by the torque T. First, calculate the shear flow in the thin wall. We have

$$q = \frac{T}{2\overline{A}} = \frac{T}{2\pi a^2}$$

Then we obtain

$$\text{Shear stress:} \qquad \tau_{xy} = \frac{q}{t} = \frac{T}{2\pi a^2 t}$$

The hoop stress σ_{yy} can be obtained by cutting the cylinder into two half-shells along the longitudinal direction and considering balance of forces. We have

$$\text{Hoop stress:} \qquad \sigma_{yy} = \frac{p_0 a}{t}$$

If the cylinder is subjected to internal pressure alone, then $N = T = 0$, and, thus $\sigma_{xx} = \tau_{xy} = 0$. The hoop stress produced by the internal pressure is

$$\sigma_{yy} = \frac{p_0}{5 \times 10^{-3}} = 200\, p_0$$

This is a uniaxial stress, and yielding is given by $\sigma_{yy} = \sigma_Y$. Thus, the value of p_0 that would cause yielding is

$$200\, p_0 = \sigma_Y$$

from which the maximum internal pressure is obtained.

$$p_0 = \frac{1}{200}\sigma_Y = 1.4 \text{ MPa}$$

If $N = 5$ MN is also present, then, in addition to the hoop stress, we have axial stress σ_{xx} given by

$$\sigma_{xx} = \frac{5 \times 10^6}{2\pi \times 1 \times 0.005} = 159 \text{ MPa}$$

At the onset of yielding, the stresses $\sigma_{xx} = 159$ MPa, $\sigma_{yy} = 200 \, p_0$, and $\tau_{xy} = 0$ must satisfy the von Mises yield criterion (6.17), i.e.,

$$\tfrac{1}{3}(159^2 + 40{,}000 \, p_0^2 - 200 \times 159 \, p_0) = \tfrac{1}{3}\sigma_Y^2$$

or, after simplification,

$$p_0^2 - 0.795 \, p_0 - 1.33 = 0$$

There are two possible solutions for the equation above, i.e.,

$$p_0 = 1.62 \text{ MPa}, \qquad -0.82 \text{ MPa}$$

We pick the first one because the second solution (-0.82 MPa) represents a pressure applied from the outside surface. Comparison of this solution with the solution for the case $N = 0$ indicates that a tensile axial force would raise the allowable internal pressure. On the other hand, it can easily be shown that axial compression would reduce the amount of internal pressure that is allowed if yielding is to be avoided.

6.3 FRACTURE MECHANICS

6.3.1 Stress Concentration

Consider a large panel with an elliptical hole as shown in Fig. 6.7a. A remote uniform normal stress σ_0 is applied. If the panel is thin, this problem can be treated as a plane stress problem. The solution for this problem, which was obtained by Inglis in 1928, can be found in many books on elasticity. Of interest is the location where the material is most severely stressed. From the solution for the stress field, we find that the normal stress σ_{yy} has the maximum value at the ends of the major axis ($x = \pm a$):

$$\sigma_{yy} = \sigma_0 \left(1 + 2\frac{a}{b}\right) \tag{6.19}$$

The distributions of σ_{yy} and σ_{xx} along the x-axis are shown in Fig. 6.7b.

The stress concentration factor K_t is defined as

$$K_t = \frac{\sigma_{yy}}{\sigma_0} = 1 + 2\frac{a}{b} \tag{6.20}$$

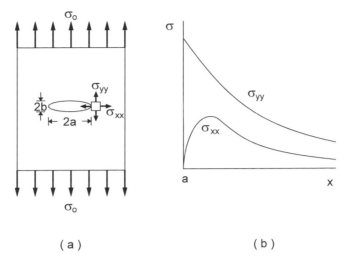

(a)   (b)

Figure 6.7 (a) Large panel with elliptical hole; (b) stress distributions along the x-axis.

For a circular hole, $a = b$ and

$$K_t = 3$$

Apparently, for elliptical holes with $a > b$, the stress concentration factor is larger than 3. In fact, theoretically, $K_t \to \infty$ as $b/a \to 0$.

The presence of stress concentrations in a material may cause premature failure. If the material contains defects such as elliptical holes, its strength definitely would depend on the magnitude of stress concentration resulting from the defects. Our ability to predict failure of the material evidently relies on our understanding of the results of these defects.

6.3.2 Concept of Cracks and Strain Energy Release Rate

Many defects and damage in materials and structures have the form of a crack. An ideal crack (or Griffith crack) can be viewed as the limiting case of an elliptic hole with the minor axis $b \to 0$. This limiting case results in unbounded stresses at the crack tip. As a result, stress-based failure criteria such as the maximum principal stress criterion and the Coulomb–Mohr criterion are not suitable for failure prediction for structures containing ideal cracks because of the singular stresses.

A. A. Griffith was the first researcher to propose an energy balance concept to determine whether a crack would grow or not. The growth of the crack signifies the onset of failure.

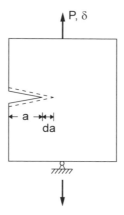

Figure 6.8   Panel with an edge crack loaded in tension.

According to Griffith's argument, new crack surfaces are formed during crack extension; the creation of crack surfaces requires a supply of energy from the system (applied forces and the material). When the supply meets the demand, crack growth is realized.

In order to put Griffith's energy concept in mathematical terms, let us consider a panel of thickness t with a crack at an edge as shown in Fig. 6.8. Assume that the crack extends the amount da after the load reaches P. During the crack extension, the load is kept at the constant value of P and thus the deflection increases by $d\delta$. In the load–deflection curve this process is represented by the line segments $\overline{AB}$ and $\overline{BC}$ in Fig. 6.9. Subsequently, the load is gradually removed as indicated by the line $\overline{CA}$. Thus, during the cycle of loading–crack extension–unloading, the energy released from the structural system is represented by the area enclosed by $ABCA$. The amount

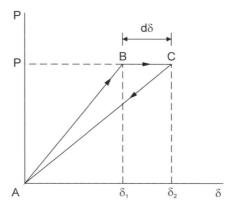

Figure 6.9   Load–deflection curve of Fig. 6.8.

of energy released, denoted by dW_s, can be expressed as

$$dW_s = \tfrac{1}{2}Pd\delta = \tfrac{1}{2}P(\delta_2 - \delta_1) \tag{6.21}$$

The energy released from the structural system (including the applied loads) provides the energy needed to form the newly created crack surfaces during the crack extension. Note that the strain energies stored in the panel before and after crack extension are given by

$$U_1 = \tfrac{1}{2}P\delta_1$$

and

$$U_2 = \tfrac{1}{2}P\delta_2$$

respectively. The increase in strain energy in the structure after crack extension is

$$dU = U_2 - U_1 = \tfrac{1}{2}P(\delta_2 - \delta_1) \tag{6.22}$$

Comparing (6.21) and (6.22), we have

$$dW_s = dU$$

This says that the amount of energy released, dW_s, during crack extension of da is equal to the gain of strain energy in the panel.

It is more convenient to quantify the energy released by normalizing with respect to the crack surface created by da, i.e.,

$$G = \frac{dW_s}{t\,da} = \frac{1}{t}\frac{dU}{da} \tag{6.23}$$

where t is the thickness of the panel and G is called the **strain energy release rate** (energy/area) per crack tip.

6.3.3 Fracture Criterion

It is our interest to determine whether the crack would grow (or extend) under a given loading condition. A crack growth criterion is derived based on the energy balance concept (the Griffith criterion) which states that if

$$G \geq G_c \tag{6.24}$$

then the crack would grow. The critical value G_c is called the **fracture toughness** of the material and is a material constant. The fracture toughness G_c can be regarded as the energy per unit area (per crack tip) needed to form fracture surfaces. If the applied load produces a G that is always larger or equal to G_c, then crack growth would not stop and catastrophic failure would occur in the structure member. The fracture criterion (6.24) is different from the strength criterion in that it assumes the presence of a crack.

Strain Energy in Structural Members   From (6.23) we note that the strain energy release rate of a crack in a structure is related to the rate of change of the total strain energy in the structure with respect to crack length. It is then useful to review some of the strain energy expressions in structural members.

Axial Element   In an axial element of a uniform cross-section, the axial stress is $\sigma_{xx} = P/A$ where P is the total applied axial force, and A is the cross-sectional area. From the strain energy density function (2.94) we have

$$W = \frac{1}{2}\sigma_{xx}\varepsilon_{xx} = \frac{1}{2}\frac{\sigma_{xx}^2}{E}$$

$$= \frac{P^2}{2\,E A^2} \quad \text{N·m/m}^3 \tag{6.25a}$$

The total strain energy stored in an axial member of length L is

$$U = \int_0^L W A\,dx = \frac{P^2 L}{2EA} \tag{6.25b}$$

Beam Element   For beams of a symmetric section subjected to a bending moment M about the y-axis, the bending stress is given by

$$\sigma_{xx} = \frac{Mz}{I}$$

The strain energy density can be expressed in the form

$$W = \frac{\sigma_{xx}^2}{2E} = \frac{M^2 z^2}{2EI^2}$$

The strain energy per unit length of the beam is obtained from integrating the strain energy density over the cross-section, i.e.,

$$\iint_A W\,dA = \frac{M^2}{2EI^2} \iint_A z^2\,dA$$

$$= \frac{M^2}{2EI} \tag{6.26}$$

Since bending moment M may be a function of x, the total strain energy stored in the beam is obtained by integrating (6.26) over the entire length. We have

$$U = \int_0^L \frac{M^2}{2EI} \, dx \tag{6.27}$$

Torsion Member   For a bar of a solid circular section, the shear stress τ is related to the torque T as

$$\tau = \frac{Tr}{J}$$

Using the shear stress-strain relation

$$\tau = G\gamma$$

the strain energy density is obtained from (2.91) as

$$W = \frac{1}{2}\tau\gamma = \frac{\tau^2}{2G} = \frac{T^2 r^2}{2GJ^2} \tag{6.28}$$

in which G is the shear modulus. The total strain energy stored in the bar of a circular cross-section and length L is

$$
\begin{aligned}
U &= \iiint_V W \, dV = \int_0^L \left(\iint_A \frac{T^2 r^2}{2GJ^2} \, dA \right) dx \\
&= \frac{T^2}{2GJ^2} \int_0^L \left(\iint_A r^2 \, dA \right) dx \\
&= \frac{T^2 L}{2GJ} \tag{6.29}
\end{aligned}
$$

For a thin-walled bar of single closed section, the strain energy stored in the bar of unit length is given by (3.47). For a bar of length L, the total strain energy is

$$U = \frac{L}{2G} \oint \frac{q^2}{t} \, ds \tag{6.30}$$

where the contour integration is along the center line of the wall.

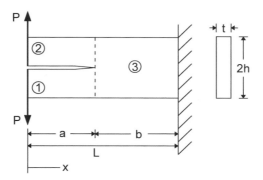

Figure 6.10   Loaded split beam.

Example 6.3   Consider an equally split beam loaded as shown in Fig. 6.10. There are three beam segments, among which segment 3 is not loaded and has no strain energy stored. Because of symmetry, beams 1 and 2 have the same amount of strain energy.

Consider beam 1. The bending moment is

$$M = Px$$

and the strain energy stored in the beam is

$$U_1 = \int_0^a \frac{M^2}{2EI} \, dx = \int_0^a \frac{P^2x^2}{2EI} \, dx = \frac{P^2a^3}{6EI}$$

where

$$I = \frac{th^3}{12}$$

is the moment of inertia of beam 1 (and beam 2). The total strain energy stored in the entire split beam is

$$U = U_1 + U_2 = 2\frac{P^2a^3}{6EI} = \frac{P^2a^3}{3EI}$$

from which we obtain the strain energy release rate

$$G = \frac{1}{t}\frac{dU}{da} = \frac{P^2a^2}{tEI}$$

Suppose that the crack would start to grow at $P = P_1$ and $a = a_1$. Thus,

$$G_c = \frac{P_1^2a_1^2}{tEI} \tag{a}$$

If $a_2 = 2a_1$, then the P_2 required to grow the crack is

$$G_c = \frac{P_2^2 a_2^2}{t E I} = \frac{4 P_2^2 (a_1)^2}{t E I} \tag{b}$$

Note that $b = L - a$ is not a constant. Comparing (a) and (b), we conclude that

$$P_2 = \tfrac{1}{2} P_1$$

That is, it takes half the load to grow the crack if the initial crack size is twice as large.

Of course, the conclusion above may not be true for other loading and structural configurations. For instance, if the transverse loads are replaced with two constant moments M_0, the strain energy release rate can readily be obtained as

$$G = \frac{M_0^2}{t E I}$$

which is obviously independent of crack length.

Example 6.4   Consider the split beam of Fig. 6.10. Change the loads to horizontal forces as shown in Fig. 6.11. In this case, segments 1 and 2 are both subjected to an axial force P, and segment 3 is subjected to a bending moment $M = Ph$ produced by the pair of axial forces. The strain energy stored in segments 1 and 2 are computed using (6.25), i.e.,

$$U_1 = \frac{P^2 a}{2 E A_1}, \qquad U_2 = \frac{P^2 a}{2 E A_2} \tag{a}$$

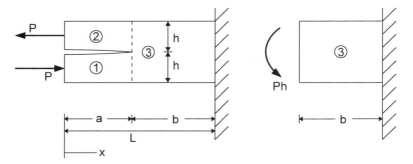

Figure 6.11   Split beam subjected to horizontal forces.

The strain energy stored in segment 3 is obtained using (6.27).

$$U_3 = \int_a^L \frac{(Ph)^2}{2EI_3} \, dx = \frac{P^2 h^2 b}{2EI_3} \tag{b}$$

The total strain energy in the split beam is

$$U = U_1 + U_2 + U_3 = \frac{P^2 a}{EA_1} + \frac{(Ph)^2 b}{2EI_3} \tag{c}$$

where $A_1 = A_2 = th$ and $I_3 = t(2h)^3/12$.

Before differentiating (c) to obtain the strain energy release rate, we should recognize the fact that $b = L - a$ is a function of a. In view of this, we have

$$G = \frac{1}{t} \frac{dU}{da} = \frac{1}{t} \left(\frac{P^2}{EA_1} - \frac{P^2 h^2}{2EI_3} \right)$$

Consider the case in which the axial forces applied to segments 1 and 2, respectively, are in the same direction. For such a loading condition, segment 3 is under an axial force of $2P$. The total strain energy in the split beam becomes

$$U = \frac{P^2 a}{EA_1} + \frac{(2P)^2 b}{2EA_3} = \frac{P^2 a}{EA_1} + \frac{2P^2(L-a)}{E(2A_1)}$$

It is easy to verify that

$$G = \frac{1}{t} \frac{dU}{da} = 0$$

This implies that the crack cannot be propagated under such loads.

6.4   STRESS INTENSITY FACTOR

Within the framework of linear elasticity, stresses near the tip of an ideal crack in a panel subjected to in-plane loads are singular. In fact, the singular stress field near the crack tip has a known functional form.

6.4.1   Symmetric Loading (Mode I Fracture)

If the loading and geometry of the cracked structure are symmetric with respect to the crack surface (e.g., Fig. 6.12), then the singular stress field (for both plane stress and plane strain) has the following form:

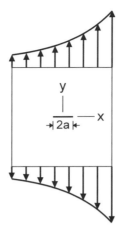

Figure 6.12   Symmetrical loading and geometry of a cracked structure.

$$\sigma_{xx} = \frac{K_I}{\sqrt{2\pi r}} \cos \frac{1}{2}\theta \left(1 - \sin \frac{1}{2}\theta \sin \frac{3}{2}\theta \right) \qquad (6.31a)$$

$$\sigma_{yy} = \frac{K_I}{\sqrt{2\pi r}} \cos \frac{1}{2}\theta \left(1 + \sin \frac{1}{2}\theta \sin \frac{3}{2}\theta \right) \qquad (6.31b)$$

$$\tau_{xy} = \frac{K_I}{\sqrt{2\pi r}} \sin \frac{\theta}{2} \cos \frac{\theta}{2} \cos \frac{3}{2}\theta \qquad (6.31c)$$

where r, θ are polar coordinates with the origin at the crack tip (see Fig. 6.13). The factor K_I is called the **stress intensity factor** for mode I fracture whose value depends on the load and geometry of the cracked structure. The unit of stress intensity factor is $Pa\sqrt{m}$.

For an *infinite* panel containing a center crack subjected to remote uniform load σ_0 (replace the load in Fig. 6.12 with a uniform load σ_0), the stress intensity factor is

$$K_I = \sigma_0 \sqrt{\pi a} \qquad (6.32)$$

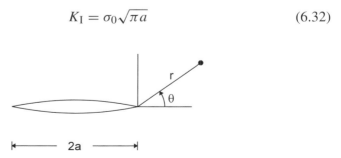

Figure 6.13   Polar coordinates with the origin at the crack tip.

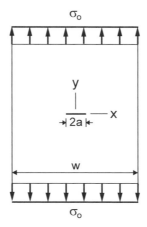

Figure 6.14   Center-cracked panel under uniform tension.

However, if the uniform remote stress σ_0 is applied in the direction parallel to the crack surfaces (i.e., in the x-direction), then a uniform state of uniaxial stress $\sigma_{xx} = \sigma_0$ is produced and there is no singular stress field near the crack tip. Consequently, $K_I = 0$.

For cracked bodies of finite dimensions under arbitrary (but symmetric) loading, stress analyses must be performed (often numerically) to obtain the stress intensity factors. For example, for a center-cracked panel of width w subjected to uniform load σ_0 (Fig. 6.14), the stress intensity factor is given by

$$K_I = Y \sigma_0 \sqrt{\pi a}$$

where Y is a width correction factor. There are many approximate correction factors for different loading conditions and geometries of cracked bodies. They can be found in many books on fracture mechanics or handbooks dedicated to stress to stress intensity factors. For a long rectangular panel with a finite width w, a few approximate width correction factors are given below.

$$Y = 1 + 0.256 \left(\frac{a}{w}\right) - 1.152 \left(\frac{a}{w}\right)^2 + 12.2 \left(\frac{a}{w}\right)^3$$

$$= \sqrt{\sec \frac{\pi a}{w}}$$

$$= \frac{1}{\sqrt{1 - (2a/w)^2}}$$

Note that, for $w \gg a$, $Y \approx 1$, and K_I reduces to that of the infinite panel.

Along the crack plane ahead of the crack tip ($\theta = 0°$), the stresses are given by

$$\sigma_{xx} = \frac{K_{\mathrm{I}}}{\sqrt{2\pi r}} \qquad (6.33a)$$

$$\sigma_{yy} = \frac{K_{\mathrm{I}}}{\sqrt{2\pi r}} \qquad (6.33b)$$

$$\tau_{xy} = 0 \qquad (6.33c)$$

Of interest is the crack opening displacement $v(x)$ along the crack surface. For the infinite panel subjected to uniform normal loading σ_0, the displacements at the upper and lower crack surfaces are symmetrical. Thus, only the displacement on the upper crack surface needs to be described:

$$v = \frac{(\kappa + 1)(1 + v)}{2E} \sigma_0 \sqrt{a^2 - x^2} \qquad (6.34)$$

where

$$\kappa = \begin{cases} 3 - 4v & \text{for plane strain} \\ \dfrac{3 - v}{1 + v} & \text{for plane stress} \end{cases} \qquad (6.35)$$

Explicitly, the plane stress crack opening displacement at the upper surface is

$$v = \frac{2}{E} \sigma_0 \sqrt{a^2 - x^2} \qquad (6.36)$$

In (6.34) and (6.36), the origin of the x-axis is located at the center of the crack as shown in Fig. 6.14.

Although the near-tip stresses given by (6.31) are not the actual stresses in the cracked structure, they are dominant near the crack tip and thus control crack growth. Also note that these near-tip stresses are in direct proportion to the stress intensity factor K_{I}. Intuitively, one would expect K_{I} to be a good parameter for determining onset of crack growth (i.e., fracture).

6.4.2 Antisymmetric Loading (Mode II Fracture)

Another mode of fracture is mode II fracture associated with loading that is antisymmetric with respect to the crack surface. Shear loading as shown in Fig. 6.15 is a mode II fracture problem. In terms of the polar coordinates shown in Fig. 6.13, the singular stress field near the crack tip is obtained as

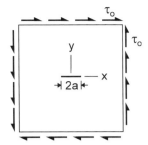

Figure 6.15   Antisymmetric shear loading.

$$\sigma_{xx} = -\frac{K_{II}}{\sqrt{2\pi r}} \sin\frac{\theta}{2}\left(2 + \cos\frac{\theta}{2}\cos\frac{3}{2}\theta\right) \tag{6.37a}$$

$$\sigma_{yy} = \frac{K_{II}}{\sqrt{2\pi r}} \sin\frac{\theta}{2}\cos\frac{\theta}{2}\cos\frac{3}{2}\theta \tag{6.37b}$$

$$\tau_{xy} = \frac{K_{II}}{\sqrt{2\pi r}} \cos\frac{\theta}{2}\left(1 - \sin\frac{\theta}{2}\sin\frac{3}{2}\theta\right) \tag{6.37c}$$

where K_{II} is the mode II stress intensity factor.

For an *infinite* cracked panel subjected to a uniform shear τ_0 as shown in Fig. 6.15, $K_{II} = \tau_0\sqrt{\pi a}$. Note that along the x-axis ahead of the crack tip ($\theta = 0°$), we have

$$\sigma_{xx} = \sigma_{yy} = 0 \tag{6.38a}$$

$$\tau_{xy} = \frac{K_{II}}{\sqrt{2\pi r}} \tag{6.38b}$$

where $r = 0$ is located at the crack tip.

The displacements on the two crack surfaces are antisymmetric with respect to the x-axis. For the upper crack surface, we have

$$u = \frac{(\kappa + 1)(1 + v)}{2E}\tau_0\sqrt{a^2 - x^2} \tag{6.39}$$

$$v = 0$$

Thus, under antisymmetric loading, the crack surfaces do not open. Rather, they slide against each other. This is why **mode II** fracture is also referred to as the **sliding mode of fracture**.

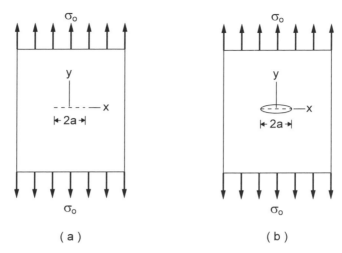

Figure 6.16 (a) Uncracked panel; (b) panel with a crack of size 2a.

6.4.3 Relation between *K* and *G*

Consider an infinite panel of unit thickness subjected to uniform tension. Figure 6.16a shows the uncracked panel, and Fig. 6.16b shows the panel with a crack of size 2a.

The total strain energy released from crack size 0 to crack size 2a can be calculated by a reverse process called the **crack closure method**. The argument is as follows. Before the crack appears, the stress field is uniform, and $\sigma_{yy} = \sigma_0$ everywhere in the panel. To return the cracked panel to its original uncracked configuration, work must be done. The amount of work necessary to close the crack is equal to the energy released during crack growth from 0 to 2a. Since the original stress is $\sigma_{yy} = \sigma_0$, to return to the original state of stress (in the uncracked panel), we need to use this stress to close the crack opening displacement v which is given by (6.34). Thus, the total energy release during crack growth from 0 to 2a is given by

$$W_s(a) = 2 \int_{-a}^{a} \tfrac{1}{2} \sigma_0 v \, dx \tag{6.40}$$

where the factor 2 accounts for the two crack surfaces. Since v is an even function, we can write

$$W_s(a) = 4 \int_{0}^{a} \tfrac{1}{2} \sigma_0 v \, dx \tag{6.41}$$

Substituting (6.34) into (6.41), we have

$$
\begin{aligned}
W_s(a) &= \frac{\sigma_0^2(\kappa+1)(1+\nu)}{E} \int_0^a \sqrt{a^2 - x^2}\, dx \\
&= \frac{\sigma_0^2(\kappa+1)(1+\nu)}{E} \left[\frac{x}{2}\sqrt{a^2 - x^2} + \frac{1}{2}a^2 \sin^{-1}\frac{x}{a} \right]_0^a \\
&= \frac{\pi \sigma_0^2 a^2 (\kappa+1)(1+\nu)}{4E}
\end{aligned}
\tag{6.42}
$$

The strain energy release rate (per crack tip) for mode I is

$$
\begin{aligned}
G_{\mathrm{I}} &= \frac{1}{2}\frac{\partial W_s}{\partial a} \\
&= \frac{\pi \sigma_0^2 a (\kappa+1)(1+\nu)}{4E} \\
&= \frac{K_{\mathrm{I}}^2 (\kappa+1)(1+\nu)}{4E}
\end{aligned}
\tag{6.43}
$$

In (6.43), the $\frac{1}{2}$ factor accounts for two crack tips in the panel.

Using the definition of κ, we can easily show that for symmetric (mode I) loading,

$$
G_{\mathrm{I}} = \frac{K_{\mathrm{I}}^2}{E} \qquad \text{for plane stress} \tag{6.44}
$$

and

$$
G_{\mathrm{I}} = \frac{1-\nu^2}{E} K_{\mathrm{I}}^2 \qquad \text{for plane strain} \tag{6.45}
$$

Following the same procedure, we can derive the strain energy release rate for antisymmetric (mode II) loading. We have

$$
G_{\mathrm{II}} = \frac{K_{\mathrm{II}}^2}{E} \qquad \text{for plane stress} \tag{6.46}
$$

and

$$
G_{\mathrm{II}} = \frac{1-\nu^2}{E} K_{\mathrm{II}}^2 \qquad \text{for plane strain} \tag{6.47}
$$

Although the relations (6.44)–(6.47) are derived for an infinite panel under either uniform normal stress σ_0 or shear stress τ_0, they are valid for finite dimensions and arbitrary loading. Of course, in general cases, K_I and K_{II} must be solved for the specific problem.

In general, it is difficult to analyze the near-tip singular stress field to obtain stress intensity factors. On the other hand, for some structures, such as beams, the energy release rate can be evaluated rather easily using the simple beam theory. The corresponding stress intensity factor can be obtained using the G–K relation. The accuracy of such an approach obviously depends on the accuracy of the simple beam theory used in calculating the strain energy. For short beams or beams of thin-walled cross-sections, the Timoshenko beam theory can provide a better description of deformation in the beam (as discussed in Chapter 4) and, thus, a more accurate strain energy release rate.

In view of these relations, the fracture toughness of a material can be given by either G_c or K_c.

Example 6.5   Consider the split beam of Example 6.3. It is obvious that the loading is a symmetric (mode I) loading. Assume a plane strain fracture condition so that

$$G_I = \frac{1 - \nu^2}{E} K_I^2$$

Since the strain energy release rate for the split beam is

$$G_I = \frac{P^2 a^2}{t E I}$$

we have

$$K_I^2 = \frac{E}{1 - \nu^2} \frac{P^2 a^2}{t E I} = \frac{P^2 a^2}{(1 - \nu^2) t I}$$

Thus,

$$K_I = P a \sqrt{\frac{1}{(1 - \nu^2) t I}}$$

Suppose that the split beam has the following dimensions:

$$h = 1 \times 10^{-2}\,\text{m}, \qquad t = 2 \times 10^{-2}\,\text{m}, \qquad a = 5 \times 10^{-2}\,\text{m}$$

If the fracture toughness of the material is

$$K_{Ic} = 24\,\text{MPa}\,\sqrt{\text{m}}$$

then the critical load P_{cr} that would cause fracture (crack extension) is obtained using the fracture criterion, i.e.,

$$K_{\rm I} = P_{\rm cr}\,a\sqrt{\frac{1}{(1-v^2)t\,I}} = K_{\rm Ic}$$

Solving the equation above for $P_{\rm cr}$ and substituting the numerical values of the beam dimensions together with $v = 0.3$, we obtain

$$P_{\rm cr} = \frac{K_{\rm Ic}}{a}\sqrt{(1-v^2)t\,I} = 2640\ {\rm N}$$

It is easy to verify that a split beam of a material with $K_{\rm Ic} = 34$ MPa $\sqrt{\rm m}$ could withstand the same load with $a = 7.08$ cm.

Example 6.6 A box beam of a rectangular thin-walled section is subjected to a counterclockwise torque as shown in Fig. 6.17a. The material is brittle,

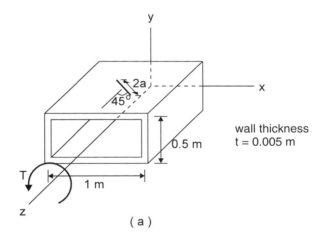

(a)

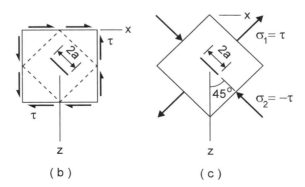

(b)　　　　(c)

Figure 6.17 Box beam of a rectangular thin-walled section subjected to torque.

and the wall material has a toughness equal to $K_{Ic} = 5$ MPa$\sqrt{m}$. After a period of service, a through-the-thickness crack appears in the top panel. The crack makes an angle $45°$ against the z-axis. Because of the crack, the torque capacity of the box beam will be reduced.

Assume $a = 0.01$ m. Cut out the 1 m $\times$ 1 m panel that contains the crack. This panel is subjected to a pure shear loading as shown in Fig. 6.17b. It is easy to show that the principal stresses are $\sigma_1 = \tau$ and $\sigma_2 = -\tau$, and the corresponding principal directions make $45°$ and $-45°$ against the x-axis as shown in Fig. 6.17c. In terms of σ_1 and σ_2 as loading, we recognize this as a mode I fracture problem.

Since σ_2 is applied parallel to the crack surface, it does not open the crack surfaces and does not contribute to the near-tip singular stresses. Hence, σ_1 is the only loading that is relevant to fracture of the panel.

We note that the size of the isolated panel (Fig. 6.17c) is large compared with the crack size. Thus, the stress intensity factor K_I can be approximated by that for an infinite panel, i.e.,

$$K_I = \sigma_1 \sqrt{\pi a} = 0.177\tau$$

The maximum shear stress τ_{max} is reached when $K_I = K_{Ic}$. We obtain

$$\tau_{max} = \frac{K_{Ic}}{0.177} = 28.2 \text{ MPa}$$

The torque capacity of the cracked box beam is

$$T = 2\overline{A}q = 2\overline{A}t\tau_{max}$$
$$= 2 \times 0.5 \times 0.005 \times 28.2 = 0.14 \text{ MN·m}$$

6.4.4 Mixed Mode Fracture

For general crack geometries and loading, both modes of fracture are present. One way to separate these modes is to separate the loading into the symmetric part and antisymmetric part. If the cracked body is symmetric with respect to the crack surface, then the symmetric part of loading produces K_I (or G_I) and the antisymmetric part, K_{II} (or G_{II}). Such decomposition is illustrated with a split beam subjected to a transverse load P as shown in Fig. 6.18.

Under mixed loading, both K_I (G_I) and K_{II} (G_{II}) contribute to fracture. There are a number of mixed mode fracture criteria available, none of which, however, stands out as the best for all materials with various properties. The following criterion has been shown to fit test data for many materials quite

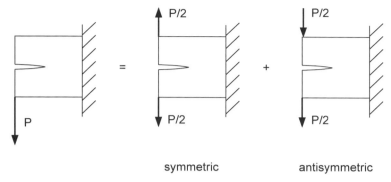

symmetric   antisymmetric

Figure 6.18   Separation of fracture models.

well:

$$\left(\frac{K_{\mathrm{I}}}{K_{\mathrm{I}c}}\right)^2 + \left(\frac{K_{\mathrm{II}}}{K_{\mathrm{II}c}}\right)^2 = 1 \tag{6.48}$$

6.5   EFFECT OF CRACK TIP PLASTICITY

A beginner in fracture mechanics is often bothered by the singular stress field near the crack tip. Of course, the stress singularity exists because of the use of linear elasticity theory in the stress analysis. In reality, the material near the crack tip yields at a finite stress. The region in which yielding occurs is called the **crack tip plastic zone**. If the plastic zone is small, then the plastic zone size has a unique relation with the stress intensity factor of the elastic singular stress field and the corresponding critical stress intensity factor K_c (fracture toughness) can still be used to characterize fracture toughness with some adjustment of the crack length. Since the value of K_c is measured experimentally, the influence of plasticity on the resistance of crack growth in the material is reflected in the value of K_c. This is a simple extension of linear elastic fracture mechanics for applications in materials of moderate ductility.

The development of the plastic zone near the crack tip depends on the yield strength and thickness of the material. In general, the stress field near the crack tip in thin plates can be approximated with a two-dimensional plane stress solution, while for sufficiently thick plates, the plane strain solution is a closer approximation. From (6.17) and (6.18) it is noted that for the same in-plane stresses σ_{xx}, σ_{yy}, and τ_{xy}, yielding conditions under plane stress can be satisfied more easily than under plane strain. This indicates that the plastic zone size near the crack tip in thin plates is greater than that

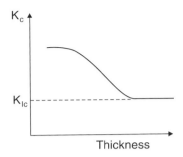

Figure 6.19   Dependence of fracture toughness on specimen thickness.

in thick plates. Since a larger plastic zone at the crack tip dissipates more energy as it moves along with the crack tip during crack growth, a larger crack tip plastic zone provides a greater resistance to crack growth. Indeed, experimental results have confirmed that thinner cracked specimens have high fracture toughnesses than thicker specimens, as depicted in Fig. 6.19. The mechanisms that cause the thickness dependency of fracture toughness are quite complicated and there is no simple physics-based model that is capable of predicting this thickness effect. In general, fracture toughness approaches a constant value as the specimen thickness reaches a certain level. This constant toughness is the **plane strain fracture toughness** of a material and is denoted by K_{Ic}. Table 6.1 lists values of K_{Ic} of some metals.

When plastic deformation occurs near the crack tip, stress becomes finite and the elastic singular stress field no longer exists. Consequently, stress intensity factor is not available for characterizing fracture toughness. New parameters are thus needed for quantifying fracture toughness for ductile

TABLE 6.1   Material properties of aluminum and steel alloys

Material	Plane Strain Toughness K_{Ic} MPa $\sqrt{\mathrm{m}}$	Yield σ_Y MPa	Ultimate Stress σ_{UT} MPa	Ultimate Elongation ε_{UT} %
Al alloys				
2024-T651	24	415	485	13
2024-T351	34	325	470	20
6061-T651	34	275	310	11
7075-T651	29	505	570	12
Steel				
AISI 1144	66	540	840	5
AISI 4130	110	1090	1150	14

Source: Data from N. E. Dowling, *Mechanical Behavior of Materials*, Prentice Hall, Englewood Cliffs, NJ, 1993, p. 282.

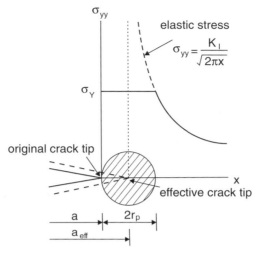

Figure 6.20   Plastic zone at the crack tip.

materials. However, for moderately ductile materials a crack length adjustment method credited to Irwin[1] renders the singular elastic stress field (the K-field) valid beyond the plastic zone. Irwin considered the elastic–plastic stress field near the crack tip as shown in Fig. 6.20. By shifting the crack tip position to a fictitious position with a distance r_p (see Fig. 6.20) and considering the balance of forces for the free body of the upper half-space above the crack, he estimated the plastic zone size $(2r_p)$ to be

$$r_p = \frac{K_I^2}{2\pi\sigma_Y^2} \tag{6.49}$$

for plane stress states. Following the same procedure, the plastic zone size under plane strain is

$$r_p = (1-2v)^2 \frac{K_I^2}{2\pi\sigma_Y^2} \tag{6.50}$$

Since the values of Poisson's ratio are around 0.3 for most metals, it is evident that the plastic zone size is much greater under plane stress than that under plane strain.

Define an effective crack length as

$$a_{\text{eff}} = a + r_p \tag{6.51}$$

[1]G. R. Irwin, J. A. Kies, and H. L. Smith, "Fracture Strengths Relative to Onset and Arrest of Crack Propagation," *Proceedings of the American Society for Testing and Materials*, Vol. 58, 1958, pp. 640–657.

Then the elastic singular stress σ_{yy} along the crack plane associated with this fictitious crack of length a_{eff} is able to describe the stress reasonably well beyond the plastic zone as long as $r_p/a \leq 0.1$.[2] Thus, the corresponding stress intensity factor K_I (a_{eff}) can be employed to characterize the opening stress σ_{yy} near the crack tip, and the critical value K_c may be used to quantify fracture toughness. Since a_{eff} depends on r_p, which in turn depends on K_I and K_I depends on a_{eff}, the calculation of K_I (a_{eff}) needs a few iterations. However, for most materials, r_p calculated from (6.49) or (6.50) using K_I based on the original crack length should be able to yield a reasonable a_{eff} and a K_I (a_{eff}).

In applying the Irwin's plastic zone adjustment method described above for determining fracture toughness, care must be exercised in making sure that the crack tip plastic zone does not interact with the boundary of the specimen. If the plastic zone gets too close to the boundary, its size and shape are affected by the boundary conditions and the K-field assumption adopted in the derivation of this method is no longer valid. This is why large specimens are often required for fracture toughness testing of materials that exhibit ductility. Without observing this restriction, the values of K_c determined with this method will depend on the specimen size.

Example 6.7   A large metallic sheet with a center crack of length $2a$ is used for testing fracture toughness K_c. The yield stress of the metal is 400 MPa, and the fracture toughness of the sheet is estimated to be about 70 MPa $\sqrt{\text{m}}$. The minimum crack length for the test needs to be determined.

The largest plastic zone size occurs when $K_I = 70$ MPa $\sqrt{\text{m}}$. Using (6.49), we have

$$r_p = 4.9 \text{ mm}$$

Thus, the half physical crack length should be at least 49 mm.

6.6 FATIGUE FAILURE

Under repeated (cyclic) loads of magnitudes below the static failure load, a structure may still fail after a number of cycles of load application. This failure is called fatigue failure, and the "duration" (or **fatigue life**) that the structure endures up to failure point is measured in terms of number of cycles of load application.

---

[2]C. T. Sun and C. Y. Wang, "A New Look at Energy Release Rate in Fracture Mechanics," *International Journal of Fracture*, Vol. 113, 2002, pp. 295–307.

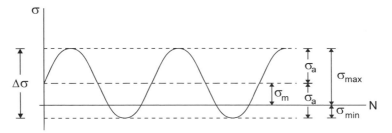

Figure 6.21   Simple cyclic loading.

6.6.1   Constant Stress Amplitude

The simplest cyclic loading is one that varies between the maximum stress σ_{max} and the minimum stress σ_{min} as shown in Fig. 6.21.

The following are terms often used in the study of fatigue behavior.

Stress range: $\quad\quad\quad\quad \Delta\sigma = \sigma_{max} - \sigma_{min}$

Stress amplitude: $\quad\quad \sigma_a \text{ (or } S_a) = \frac{1}{2}\Delta\sigma$

Mean stress: $\quad\quad\quad \sigma_m = \frac{1}{2}(\sigma_{max} + \sigma_{min})$

Stress ratio: $\quad\quad\quad R = \sigma_{min}/\sigma_{max}$

Fatigue life: $\quad\quad\quad N_f = $ number of cycles to failure

Note that the quantities above are related, i.e.,

$$\sigma_{max} = \sigma_m + \sigma_a$$

$$\sigma_{min} = \sigma_m - \sigma_a$$

$$\Delta\sigma = \sigma_{max}(1 - R)$$

$$\sigma_m = \frac{1}{2}\sigma_{max}(1 + R)$$

Thus, the cyclic load can be given in various ways. For example, the loading as depicted in Fig. 6.21 can be given by specifying σ_{max} and σ_{min}, σ_a and σ_m, or σ_{max} and R, and so on.

6.6.2   *S–N* Curves

Constant amplitude (with different mean stresses) fatigue tests are often used to evaluate the fatigue properties of a material. The results are plotted in stress amplitude (σ_a or S_a) versus cycles to failure N_f. This is the ***S–N* curve**. A

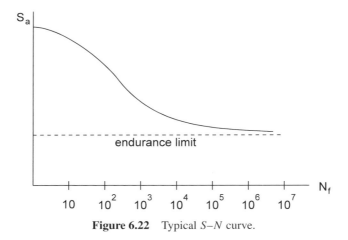

Figure 6.22   Typical $S-N$ curve.

typical $S-N$ curve is shown in Fig. 6.22. Similar curves can be obtained by plotting σ_{max} or $\Delta\sigma$ versus N_f.

For some materials, there may exist a stress amplitude below which fatigue failure does not appear ever to occur. This stress level is called the **fatigue limit** or **endurance limit**. If a structure is designed to last for a specified number of cycles (N_f), the $S-N$ curve can provide the allowable stress.

6.6.3   Variable Amplitude Loading

Consider a fatigue loading consisting of m stress amplitudes, say, $\sigma_{a1}, \sigma_{a2}, \ldots, \sigma_{am}$. For each stress amplitude σ_{ai} the number of loading cycles is N_i. The **Palmgren–Miner rule** is often used to estimate fatigue life under such loading. The Palmgren–Miner rule is based on the assumption that for each stress amplitude σ_{ai} applied for N_i cycles, the total fatigue life is depleted by the amount (percentage of total fatigue life under σ_{ai})

$$\frac{N_i}{N_{fi}}$$

Fatigue failure occurs if 100 percent of the fatigue life is consumed, i.e.,

$$\frac{N_1}{N_{f1}} + \frac{N_2}{N_{f2}} + \cdots = 1$$

or

$$\sum_{i=1}^{m} \frac{N_i}{N_{fi}} = 1 \tag{6.52}$$

Mainly because of its simplicity, Palmgren–Miner's rule has been widely used for estimating fatigue life of structures under variable amplitude loading. However, experimental results indicate that this formula often produces predictions that are not accurate and are not necessarily on the conservative side.

6.7 FATIGUE CRACK GROWTH

Fatigue damage often appears in the form of cracks. If the crack reaches a critical length at which the stress intensity factor or the strain energy release rate is equal to the critical value K_c or G_c (the fracture toughness), then catastrophic failure would occur. Fatigue damage can be considered to have two parts, namely, the crack initiation part (formation of cracks) and the crack propagation part. Accordingly, fatigue life is composed of the number of cycles of load application until a detectable size of crack appears and the number of cycles of subsequent load application until the crack grows to the critical length.

Another approach to fatigue failure is to assume the presence of flaws (cracks) in the structure before it is placed in service. Such flaws could be produced during manufacture of the material, forming, or machining of the structural part. With a realistically assumed flaw (crack) size in the structure at the most critical location, fatigue life can be estimated purely based on crack growth. We are interested in determining the crack growth per cycle of load application, da/dN, under mode I loading. Since the singular stress field at the crack tip is proportional to the stress intensity factor [see (6.31)], it is reasonable to assume that the crack growth rate (da/dN) is a function of the stress intensity factor K_I or the range of stress intensity factor ΔK_I in the case of cyclic loading, i.e.,

$$\frac{da}{dN} = f(\Delta K), \qquad \Delta K = K_{max} - K_{min} \qquad (6.53)$$

where, for simplicity, K is used to denote K_I. A typical experimentally obtained crack growth curve is shown in Fig. 6.23 in a log-log plot. This indicates that the power law

$$\frac{da}{dN} = C(\Delta K)^m \qquad (6.54)$$

fits the data very well except for the two ends of the ΔK range. This crack growth equation is credited to P. C. Paris and is usually called the **Paris Law** (or more correctly, the Paris model) of crack growth.

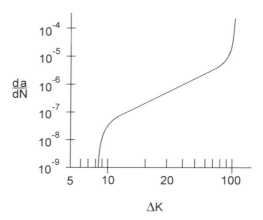

Figure 6.23   Typical experimentally obtained crack growth curve.

In log-log scale, (6.54) can be expressed as

$$\log \frac{da}{dN} = \log C + m \log(\Delta K) \qquad (6.55)$$

Thus, in the log-log plot, the relation between da/dN and ΔK appears linear.

Equation (6.54) can be integrated to find the relation between crack length and load cycles. To do this, the relation between K and crack length a must be obtained first. For example, consider a large panel (treated as an infinite panel) containing initially a center crack of $2a_0$. If the cyclic loading is uniform normal stress with

$$\sigma_{\max} = \sigma_0$$
$$\sigma_{\min} = 0$$

then

$$\Delta K = \sigma_0 \sqrt{\pi a}$$

From (6.54), we have

$$\frac{da}{dN} = C(\sigma_0 \sqrt{\pi a})^m$$

or

$$\frac{da}{a^{m/2}} = C(\sigma_0^2 \pi)^{m/2} \, dN$$

TABLE 6.2 Values of C and m for K in MPa$\sqrt{\text{m}}$ and da/dN in m/cycle

Material	C	m
2024T3 ($R = 0.1$)	1.60×10^{-11}	3.59
2024T3 ($R = 0.5$)	3.15×10^{-11}	3.59
Martensitic steel ($R = 0$)	1.36×10^{-10}	2.25
Austenitic steel ($R = 0$)	5.60×10^{-12}	3.25

Integrating the above equation, i.e.,

$$\int_{a_0}^{a} \frac{da}{a^{m/2}} = \int_{0}^{N} C(\sigma_0^2 \pi)^{m/2} \, dN$$

we obtain

$$\frac{1}{-m/2 + 1} \left(a^{-m/2+1} - a_0^{-m/2+1} \right) = C(\sigma_0^2 \pi)^{m/2} N$$

Thus, the current half crack length after N cycles is

$$a = \left[\left(-\frac{m}{2} + 1 \right) C(\sigma_0^2 \pi)^{m/2} N + a_0^{-m/2+1} \right]^{1/(-m/2+1)}$$

In the Paris model, coefficients C and m are independent of ΔK (or equivalently, the stress range) but are influenced by the R-ratio. Table 6.2 lists values of C and m for some metals.

PROBLEMS

6.1 Derive the distortional energy expression for plane stress.

6.2 A thin-walled hollow sphere 2 m in diameter is subjected to internal pressure p_0. The wall thickness is 5 mm and the yield stress of the material is 250 MPa. Use both Tresca and von Mises yield criteria to determine the maximum internal pressure p_0 that does not cause yielding.

6.3 Consider the problem of Example 6.2. Find the maximum p_0 without causing yielding if $N = -50 \times 10^6$ N (compression).

6.4 An aluminum alloy 2024-T651 (see Table 6.1) panel is subjected to biaxial loading as shown in Fig. 6.24. Assume that $\sigma_1 = 300$ MPa and σ_2 can be either tension or compression. Find the maximum values of $|\sigma_2|$ in tension and compression that the panel can withstand before yielding according to von Mises yield criterion.

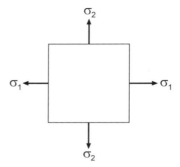

Figure 6.24   Material under biaxial stress.

6.5 Find the total strain energy release rate for the split beam loaded as shown in Figs. 6.25 and 6.26.

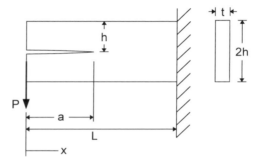

Figure 6.25   Split beam subjected to shear force.

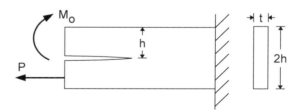

Figure 6.26   Split beam subjected to extension and bending.

6.6 Consider the split beam with loading shown in Fig. 6.27. Loadings in both Figs. 6.11 and 6.27 are antisymmetric, and both are mode II fracture problems. For the same value of P, which loading is more efficient in cracking the beam? Assume that the beam dimensions and the elastic properties are

$$E = 70 \, \text{GPa}, \qquad \nu = 0.3$$
$$a = 10 \times 10^{-2} \, \text{m}, \qquad t = 2 \times 10^{-2} \, \text{m}$$
$$L = 15 \times 10^{-2} \, \text{m}, \qquad h = 1 \times 10^{-2} \, \text{m}$$

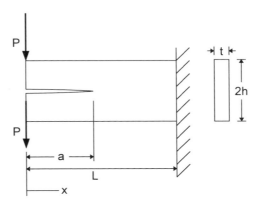

Figure 6.27   Split beam subjected to shear forces.

6.7 To further split a beam, a rigid pin of diameter $d = 0.5$ cm is inserted as shown in Fig. 6.28. How far does one have to drive the cylinder in order to split the beam? Assume a plane strain fracture condition with $K_{Ic} = 50 \, \text{MPa} \, \sqrt{\text{m}}$.

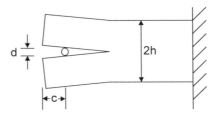

Figure 6.28   Split beam opened by a cylinder.

6.8 Consider a long thin-walled cylinder of a brittle material subjected to an internal pressure p_0. The diameter of the cylinder is 2 m, the wall thickness is 5 mm, and the mode I fracture toughness of the material (of the same thickness of the wall) is $K_{Ic} = 5 \, \text{MPa} \, \sqrt{\text{m}}$ (here, K_{Ic} may not be the plane strain fracture toughness). If there is a through-the-thickness longitudinal crack of 5 cm in length on the cylinder, estimate the maximum internal pressure that the cracked cylinder can withstand. If the cracked cylinder is subjected to a torque and the mode II toughness of the material is the same as that of mode I, estimate the maximum torque. Provide justifications of the approach employed in the estimation.

6.9 Consider the thin-walled box beam in Fig. 6.17. The top wall contains a crack parallel to the x-axis. The crack length is 0.02 m (i.e., $a = 0.01$ m). Assume that the material is brittle and that modes I and II have the same toughness value of 5 MPa $\sqrt{m}$. If the box beam has already been subjected to a torque $T = 100$ kN $\cdot$ m, estimate the maximum additional axial force N by using the mixed mode fracture criterion.

6.10 Derive the strain energy (bending and shear together) per unit length of a Timoshenko beam with a solid rectangular cross-section. The counterpart of the simple beam theory is given by (6.26). Use this expression to derive the mode I strain energy release rate for the split beam of Fig. 6.10. Compare the Timoshenko beam solution with the simple beam solution. How long (in terms of a/h) does the crack length have to be for the simple beam solution to be within 5 percent of the Timoshenko beam solution?

6.11 Compare the plastic zone sizes for plane strain mode I fracture at failure in Al 2024-T651 and Al 7075-T651.

6.12 A center-cracked thin Al 2024-T651 flat panel with a very large width-to-crack length ratio is subjected to uniform remote tensile stress. The initial crack length is 50 mm and it grows to 55 mm when the applied load reaches the maximum value of 136 MPa. Determine the fracture toughness using Irwin's plastic zone adjustment method. Is the crack length valid for this method?

6.13 The split beam of Fig. 6.10 is subjected to a pair of cyclic opening forces P with

$$P_{max} = 2000 \text{ N}, \qquad P_{min} = 0$$

The initial crack length a_0 is 40 mm. The material is 2024-T651 Al, and $t = 2 \times 10^{-2}$ m, $h = 1 \times 10^{-2}$ m. The crack growth rate is given by

$$\frac{da}{dN} = 1.6 \times 10^{-11} (\Delta K_I)^{3.59} \text{ m/cycle}$$

in which K_I is in MPa$\sqrt{m}$. Find the number of cycles to failure (at which the crack becomes unstable under the load P_{max}).

6.14 Consider Example 6.6. Instead of a static torque, a cyclic torque with

$$T_{max} = 0.1 \text{ MN·m}, \qquad T_{min} = 0$$

is applied. The Paris law for the material is

$$\frac{da}{dN} = 5 \times 10^{-11}(\Delta K_{\mathrm{I}})^3 \qquad \text{m/cycle}$$

Find the number of cycles for the crack of initial length $a_0 = 0.01$ m to grow to a length $a = 0.02$ m.

7

ELASTIC BUCKLING

Failure in a structure can be classified in two general categories, material failure and structural failure. The former includes plastic yielding, rupture, fatigue, and unstable crack growth (fracture). In the latter category examples include flutter (excessive dynamic deflection of a structure in airflow) and buckling. Structural failure results in the loss of the designed structural functions and may lead to eventual material failure. In this chapter we study an important mode of structural failure—buckling. For the sake of brevity, the transverse displacements of the bar are denoted by w and v instead of the w_0 and v_0 used in Chapter 4.

7.1 ECCENTRICALLY LOADED BEAM-COLUMN

A bar is called a **beam-column** if it is subjected to bending moments and compressive axial forces. An example is the cantilever bar subjected to an eccentrical load as shown in Fig. 7.1. The vertical arm attached to the free end is assumed to be rigid.

Referring to the deformed configuration shown in Fig. 7.1, the bending moment at section x is

$$M = -P(\delta + e - w)$$

where $w(x)$ is the transverse deflection of the bar and δ is the deflection at the free end. Using the bending equation (4.14), we have

$$EI\frac{d^2w}{dx^2} = -M = P(\delta + e - w)$$

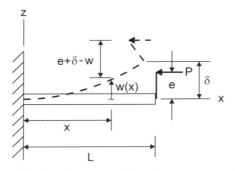

Figure 7.1 Cantilever bar subjected to an eccentric load.

which can be expressed as

$$\frac{d^2w}{dx^2} + k^2w = k^2(\delta + e) \tag{7.1}$$

where

$$k = \sqrt{\frac{P}{EI}} \tag{7.2}$$

The general solution for the nonhomogeneous linear ordinary differential equation (7.1) is easily obtained as

$$w(x) = C_1 \sin kx + C_2 \cos kx + \delta + e \tag{7.3}$$

The three unknowns C_1, C_2, and δ are determined using the boundary conditions, i.e.,

$$\text{at } x = 0, \ w = 0: \quad C_2 = -\delta - e \tag{7.4}$$

$$\frac{dw}{dx} = 0: \quad C_1 = 0 \tag{7.5}$$

and

$$\text{at } x = L, w = \delta: \quad C_1 \sin kL + C_2 \cos kL + \delta + e = \delta \tag{7.6}$$

Solving the three equations above, we obtain

$$C_2 = -\frac{e}{\cos kL}, \qquad C_1 = 0 \tag{7.7}$$

$$\delta = e\left(\frac{1}{\cos kL} - 1\right) \tag{7.8}$$

with which we have the deflection as

$$w(x) = e \left(\frac{1 - \cos kx}{\cos kL} \right) \tag{7.9}$$

From (7.8) we see that the tip deflection δ is proportional to the eccentricity e, and that $\delta \to \infty$ as $kL \to \pi/2$ no matter how small $e(\neq 0)$ is. The compressive force corresponding to $kL = \pi/2$ is

$$P_{\mathrm{cr}} = \frac{\pi^2 E I}{4 L^2} \tag{7.10}$$

At this load, a straight bar under axial compression would suffer excessive deflection (or buckling). This critical load as given by (7.10), which is independent of the eccentricity e, is called the **buckling load** for a "straight" bar with one end clamped and the other free.

7.2 ELASTIC BUCKLING OF STRAIGHT BARS

For the eccentrically compressed bar discussed in Section 7.1, the deflection given by (7.9) vanishes if $e = 0$ and $P < P_{\mathrm{cr}}$. That is, no transverse deflection can be produced by the compressive force. However, when $e = 0$ and $P = P_{\mathrm{cr}}$ (or equivalently, $\cos kL = 0$), the solution (7.9) for the deflection is not determined. Hence, to consider the axially loaded ($e = 0$) bar, a different formulation of the problem is needed. Consider a centrically ($e = 0$) compressed straight bar. The boundary conditions at the two ends are arbitrary. We want to examine for a given P whether it is possible to maintain a transverse deflection $w(x)$. If such deflection is possible, then it must satisfy the equilibrium equation and the specified boundary conditions.

Take a differential element from the compressed bar in the (assumed) buckled position as shown in Fig.7.2. The equilibrium equations for this free

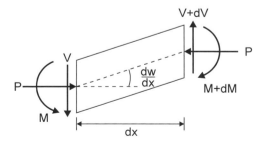

Figure 7.2   Compressed bar in the buckled position.

body are

$$\sum F_z = 0: \quad (V + dV) - V = 0$$

or

$$\frac{dV}{dx} = 0 \tag{7.11}$$

$$\sum M = 0: \quad (V + dV)\, dx + P\frac{dw}{dx}\, dx = dM$$

or

$$V = \frac{dM}{dx} - P\frac{dw}{dx} \tag{7.12}$$

Substituting (7.12) into (7.11), we obtain the equilibrium equation for the assumed deflection

$$\frac{d^2 M}{dx^2} - P\frac{d^2 w}{dx^2} = 0 \tag{7.13}$$

Substituting the relation

$$M = -EI\frac{d^2 w}{dx^2} \tag{7.14}$$

into (7.13), we obtain the equilibrium equation in terms of deflection as

$$\frac{d^4 w}{dx^4} + k^2\frac{d^2 w}{dx^2} = 0 \tag{7.15}$$

where k is defined by (7.2). The general solution for (7.15) is readily obtained as

$$w = C_1 \sin kx + C_2 \cos kx + C_3 x + C_4 \tag{7.16}$$

The four arbitrary constants C_1 to C_4 are to be determined by using the boundary conditions.

7.2.1 Pinned–Pinned Bar

Consider a straight bar with pinned ends as shown in Fig. 7.3. The boundary conditions are:

$$\text{at } x = 0: \quad w = 0, \tag{7.17a}$$

$$M = -EI\frac{d^2 w}{dx^2} = 0 \tag{7.17b}$$

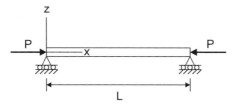

Figure 7.3   Straight bar with pinned ends.

and

$$\text{at } x = L: \quad w = 0, \tag{7.17c}$$

$$M = -EI\frac{d^2w}{dx^2} = 0 \tag{7.17d}$$

Substitution of (7.16) into (7.17) yields

$$C_2 + C_4 = 0 \tag{7.18a}$$

$$C_2 = 0 \tag{7.18b}$$

$$C_1 \sin kL + C_2 \cos kL + C_3 L + C_4 = 0 \tag{7.18c}$$

$$C_1 k^2 \sin kL + C_2 k^2 \cos kL = 0 \tag{7.18d}$$

From (7.18a) and (7.18b) we have $C_2 = C_4 = 0$. Thus, the four equations reduce to

$$C_1 \sin kL + C_3 L = 0 \tag{7.19a}$$

$$C_1 \sin kL = 0 \tag{7.19b}$$

from which we obtain $C_3 = 0$ and

$$C_1 \sin kL = 0 \tag{7.20}$$

Since $C_1 \neq 0$ (otherwise, we have a trivial solution, i.e., $w = 0$ everywhere), we must require that

$$\sin kL = 0 \tag{7.21}$$

Equation (7.21) is satisfied if

$$kL = n\pi, \quad n = 1, 2, 3, \dots \tag{7.22}$$

The corresponding P's that satisfy (7.22) are

$$P_{\text{cr}}^{(n)} = \frac{n^2\pi^2 EI}{L^2}, \quad n = 1, 2, 3, \dots \tag{7.23}$$

The deflection (**buckling mode shape**) for each **critical load** $P_{cr}^{(n)}$ (also called **buckling load**) is

$$w^{(n)}(x) = C_1 \sin k^{(n)} x \tag{7.24}$$

where

$$k^{(n)} = \sqrt{\frac{P_{cr}^{(n)}}{EI}} \tag{7.25}$$

Hence, there are infinitely many possible deformed configurations given by (7.24) that are associated with the axial loads given by (7.23). In other words, at the compressive load $P_{cr}^{(n)}$, besides the straight position, the bar can also assume a deformed position given by (7.24) for any value of C_1. These values of P, the critical loads, are also called **bifurcation points**.

Among all the critical loads, the lowest one with $n = 1$ is of particular importance because as compression is applied to the bar the lowest buckling load is reached first. The lowest buckling load (for the first buckling mode with $n = 1$) is

$$P_{cr} = P_{cr}^{(1)} = \frac{\pi^2 EI}{L^2} \tag{7.26}$$

which is known as **Euler's formula** for column buckling. The corresponding mode shape is

$$w(x) = C_1 \sin \frac{\pi x}{L} \tag{7.27}$$

To produce the second buckling mode, the first mode must be suppressed. This can be realized by adding a support at the midspan of the bar as shown in Fig. 7.4. In view of the buckling mode shape (7.27), we must set $C_1 = 0$ for the first mode to satisfy the condition $w = 0$ at $x = L/2$. Thus, the first buckling mode is suppressed.

Consider the second buckling mode ($n = 2$), for which the buckling load is

$$P_{cr}^{(2)} = \frac{4\pi^2 EI}{L^2} \tag{7.28}$$

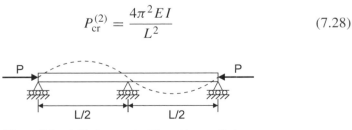

Figure 7.4   Added support at the midspan of a bar.

and the corresponding buckling mode shape is given by

$$w(x) = C_1 \sin \frac{2\pi x}{L} \tag{7.29}$$

which satisfies the condition $w = 0$ at the midspan. Thus, the lowest compressive load that can cause buckling of the bar of Fig. 7.4 is that given by (7.28). This load is four times the buckling load for the bar without support at the midspan.

The discussion above indicates that the buckling load of a bar can be increased significantly by reducing the span between supports. This is a common practice in aircraft structures to increase the compressive load capability of stringers without changing their bending stiffness by decreasing the spacing between adjacent ribs or frames.

7.2.2 Clamped–Free Bar

The clamped–free bar shown in Fig. 7.1 is assumed to be centrically compressed ($e = 0$). The boundary conditions are

$$\text{at } x = 0: \qquad w = 0 \tag{7.30a}$$

$$\frac{dw}{dx} = 0 \tag{7.30b}$$

$$\text{at } x = L: \qquad M = -EI\frac{d^2 w}{dx^2} = 0 \tag{7.30c}$$

$$V = \frac{dM}{dx} - P\frac{dw}{dx} = 0$$

The last equation can be written as

$$\frac{d^3 w}{dx^3} + k^2 \frac{dw}{dx} = 0 \tag{7.30d}$$

Substituting the general solution (7.16) into (7.30), we obtain

$$C_2 + C_4 = 0 \tag{7.31a}$$

$$C_1 k + C_3 = 0 \tag{7.31b}$$

$$C_1 k^2 \sin kL + C_2 k^2 \cos kL = 0 \tag{7.31c}$$

$$C_3 k^2 = 0 \tag{7.31d}$$

From (7.31b) and (7.31d) we have $C_1 = C_3 = 0$. From (7.31c) we obtain

$$C_2 \cos kL = 0 \tag{7.32}$$

For a nontrivial solution $C_2 \neq 0$, and we require that

$$\cos kL = 0 \quad \text{or} \quad k = \frac{n\pi}{2L}, \qquad n = 1, 3, 5, \ldots \tag{7.33}$$

which yields the buckling loads:

$$P_{cr}^{(n)} = \frac{n^2\pi^2 EI}{4L^2}, \qquad n = 1, 3, 5, \ldots \tag{7.34}$$

The lowest buckling load is

$$P_{cr} = \frac{\pi^2 EI}{4L^2} \tag{7.35}$$

which is only one-fourth that for the pinned–pinned bar.

The lowest buckling mode shape is

$$w(x) = C_2 \cos kx + C_4 = C_2(\cos kx - 1) \tag{7.36}$$

which is identical to that of the eccentrically loaded bar (see Fig. 7.1) except for the constant amplitude.

7.2.3 Clamped–Pinned Bar

The boundary conditions for the clamped–pinned ends (see Fig. 7.5) are

$$\text{at } x = 0: \quad w = 0, \quad \frac{dw}{dx} = 0 \tag{7.37a}$$

$$\text{at } x = L: \quad w = 0, \quad \frac{d^2w}{dx^2} = 0 \tag{7.37b}$$

which yield the following four equations:

$$C_2 + C_4 = 0 \tag{7.38a}$$

$$C_1 k + C_3 = 0 \tag{7.38b}$$

$$C_1 \sin kL + C_2 \cos kL + C_3 L + C_4 = 0 \tag{7.38c}$$

$$C_1 \sin kL + C_2 \cos kL = 0 \tag{7.38d}$$

Eliminating C_3 and C_4 from (7.38c) and (7.38d) using (7.38a) and (7.38b), we obtain

$$C_1(\sin kL - kL) + C_2(\cos kL - 1) = 0 \tag{7.39a}$$

$$C_1 \sin kL + C_2 \cos kL = 0 \tag{7.39b}$$

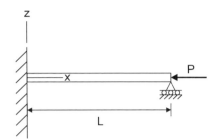

Figure 7.5   Bar with clamped–pinned ends.

It is easy to verify that neither $\sin kL = 0$ nor $\cos kL = 0$ can satisfy (7.39) simultaneously.

From (7.39b), we have

$$C_2 = -\frac{\sin kL}{\cos kL} C_1 \tag{7.40}$$

Substituting (7.40) into (7.39a) yields

$$C_1 (\tan kL - kL) = 0 \tag{7.41}$$

For a nontrivial solution, we require that

$$\tan kL - kL = 0 \tag{7.42}$$

The solution for kL to (7.42) can only be solved numerically. The lowest value that satisfies (7.42) is approximately

$$kL = 4.49 \tag{7.43}$$

from which the lowest buckling load is obtained as

$$P_{cr} = \frac{(4.49)^2 EI}{L^2} = \frac{20.16 EI}{L^2} \tag{7.44}$$

From (7.40), (7.38a), and (7.38b), we have

$$C_2 = -4.49\,C_1$$

$$C_3 = -kC_1 = -\frac{4.49}{L} C_1$$

$$C_4 = 4.49\,C_1$$

Thus, the buckling shape is not a pure sinusoid. The buckling load is about twice that for the pinned–pinned bar.

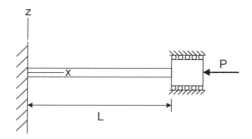

Figure 7.6   Clamped–clamped bar.

7.2.4   Clamped–Clamped Bar

The boundary (end) conditions for the clamped–clamped bar (Fig. 7.6) are $w = 0$, and $dw/dx = 0$ at both ends ($x = 0, L$). Using the general solution (7.16), these boundary conditions become

$$C_2 + C_4 = 0 \tag{7.45a}$$

$$C_1 k + C_3 = 0 \tag{7.45b}$$

$$C_1 \sin kL + C_2 \cos kL + C_3 L + C_4 = 0 \tag{7.45c}$$

$$C_1 k \cos kL - C_2 k \sin kL + C_3 = 0 \tag{7.45d}$$

Eliminating C_3 and C_4 from (7.45c) and (7.45d) using (7.45a) and (7.45b), we obtain

$$C_1 (\sin kL - kL) + C_2 (\cos kL - 1) = 0 \tag{7.46a}$$

$$C_1 (\cos kL - 1) - C_2 \sin kL = 0 \tag{7.46b}$$

Consider the possible solution

$$\sin kL = 0, \qquad kL = m\pi, \qquad m = 1, 2, 3, \ldots \tag{7.47}$$

For $m = 1, 3, 5, \ldots$, we have

$$\cos kL = -1 \tag{7.48}$$

From (7.46a) and (7.46b), the conditions (7.47) and (7.48) require that $C_1 = 0$ and $C_2 = 0$. Thus, we have a trivial solution.

For $m = 2, 4, 6, \ldots$ (or $m = 2n$, $n = 1, 2, 3, \ldots$),we have

$$\sin kL = 0 \quad \text{and} \quad \cos kL = 1 \tag{7.49}$$

Substitution of (7.49) into (7.46) yields $C_1 = 0$ (thus, $C_3 = 0$) and $C_2 \neq 0$. Thus, the buckling mode shape is

$$w(x) = C_2 \cos kx + C_4 = C_2(\cos kx - 1) \tag{7.50}$$

with

$$k = \frac{2n\pi}{L}, \qquad n = 1, 2, 3, \ldots$$

Thus, the buckling loads are

$$P_{\text{cr}}^{(n)} = \frac{4n^2\pi^2 EI}{L^2}, \qquad n = 1, 2, 3, \ldots \tag{7.51}$$

The lowest buckling load

$$P_{\text{cr}} = \frac{4\pi^2 EI}{L^2} \tag{7.52}$$

is four times the buckling load for the pinned–pinned bar.

7.2.5   Effective Length of Buckling

One common feature among all buckling loads is that they can be expressed in a single form as

$$P_{\text{cr}} = \frac{\pi^2 EI}{L_e^2} \tag{7.53}$$

where L_e is the **effective length of buckling** whose value depends on the boundary conditions. For example, for the pinned–pinned bar, $L_e = L$; for the clamped–pinned bar, $L_e = 0.7L$; for the clamped–free bar, $L_e = 2L$; and for the clamped–clamped bar $L_e = 0.5L$.

The buckling load is directly proportional to the bending stiffness and inversely proportional to the square of the effective length. It is thus more efficient to increase the buckling strength of a bar by reducing its effective length of buckling.

Since the moment of inertia I of the bar cross-section is, like L_e, a geometric quantity, (7.53) is often expressed as

$$P_{\text{cr}} = \frac{\pi^2 EA}{(L_e/\rho)^2} \tag{7.54}$$

where ρ is the radius of gyration defined by

$$I = \rho^2 A \quad (A = \text{cross-sectional area}) \tag{7.55}$$

The quantity L_e/ρ is referred to as the **effective slenderness ratio** of the bar.

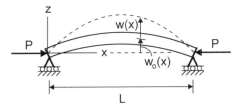

Figure 7.7 Initially deformed bar.

7.3 INITIAL IMPERFECTION

In Section 7.2, the bar was assumed to be initially straight. If the bar is slightly bent, possibly due to forced fitting, then the equilibrium equation (7.12) must be modified. Assume that the initially deformed shape of the bar is given by $w_0(x)$, (see Fig. 7.7) and the additional perturbed deflection is denoted by $w(x)$. Then the total deflection is given by

$$\overline{w}(x) = w(x) + w_0(x) \tag{7.56}$$

The equilibrium equation (7.12) becomes

$$V = \frac{dM}{dx} - P\frac{d\overline{w}}{dx} = \frac{dM}{dx} - P\left(\frac{dw}{dx} + \frac{dw_0}{dx}\right) \tag{7.57}$$

This leads to the equilibrium equation for the bent bar as

$$\frac{d^4w}{dx^4} + k^2\frac{d^2w}{dx^2} = -k^2\frac{d^2w_0}{dx^2} \tag{7.58}$$

The complementary solution for (7.58) is given by (7.16). The particular solution requires a specific form of w_0.

For illustrative purposes, consider a pinned–pinned bar with the initial bent shape given by

$$w_0 = \delta_0 \sin\frac{\pi x}{L} \tag{7.59}$$

Substituting (7.59) into (7.58), the equilibrium equation becomes

$$\frac{d^4w}{dx^4} + k^2\frac{d^2w}{dx^2} = \delta_0 k^2\frac{\pi^2}{L^2}\sin\frac{\pi x}{L} \tag{7.60}$$

It is easy to verify that

$$w_p = \frac{k^2\delta_0}{(\pi^2/L^2) - k^2}\sin\frac{\pi x}{L} \tag{7.61}$$

is a particular solution. The complete solution is the sum of the complementary and the particular solutions, i.e.,

$$w = C_1 \sin kx + C_2 \cos kx + C_3 x + C_4 + w_p \qquad (7.62)$$

The boundary conditions yield four homogeneous equations identical to (7.18a)–(7.18d) which can be satisfied by $C_1 = C_2 = C_3 = C_4 = 0$. Thus, the solution (7.62) reduces to

$$w = w_p = \frac{k^2 \delta_0}{(\pi^2/L^2) - k^2} \sin \frac{\pi x}{L} \qquad (7.63)$$

for which the maximum deflection $w(L/2)$ is

$$w\left(\frac{L}{2}\right) = \frac{k^2 \delta_0}{(\pi^2/L^2) - k^2} = \frac{P\delta_0}{(\pi^2 EI/L^2) - P} = \frac{P\delta_0}{P_{cr} - P}$$
$$= \frac{\delta_0}{(P_{cr}/P) - 1} \qquad (7.64)$$

In (7.64), $P_{cr} = \pi^2 EI/L^2$ is the lowest buckling load of the straight pinned–pinned bar.

Figure 7.8 depicts the maximum deflection $w(L/2)$ as a function of P/P_{cr} for various values of the initial deflection δ_0. The deflection becomes unbounded as P approaches the buckling load P_{cr}. Thus, the amount of initial imperfection of the bar influences only the amplitude of the deflection but not the limiting (buckling) load.

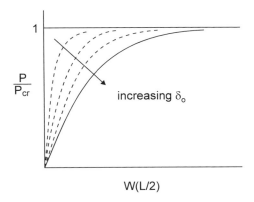

Figure 7.8   Load–deflection curves for different initial imperfections.

7.4   POSTBUCKLING BEHAVIOR

In the previous sections, we found that when a bar is subjected to the buckling load, its deflection would become unbounded. This is the result of using small deflection beam theory. In reality, the compressive load can be increased beyond the buckling load if the large deflection effect is included in the formulation of the equilibrium equation.

For a small deflection, we have used the approximate relation

$$M = -EI\frac{d^2w}{dx^2} \tag{7.65}$$

If the exact curvature $d\theta/ds$ is used, the relation above becomes

$$M = -EI\frac{d\theta}{ds} \tag{7.66}$$

where s is the contour along the deformed beam and θ is the slope of the deflection curve (see Fig. 7.9).

If the deflection is very small so that the slope θ is very small and $ds \simeq dx$, then we adopt the approximate expressions

$$\frac{dw}{dx} = \theta, \qquad \frac{d\theta}{ds} = \frac{d\theta}{dx} \tag{7.67}$$

and thus the relation given by (7.65). For a beam undergoing large deflection, (7.66) must be used.

Consider a buckled bar with pinned–pinned end conditions as shown in Fig. 7.10. Take the contour s to coincide with the deformed bar which has a length of L. The initial slope of the deflection at the left end is denoted by α. From Fig. 7.10, it is easy to see that the bending moment along the deformed

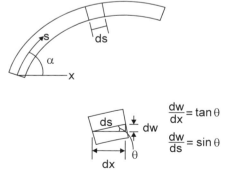

Figure 7.9   Bar in large deflection.

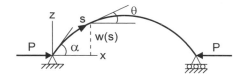

Figure 7.10 Buckled bar with a pinned–pinned end.

bar is

$$M(s) = Pw(s) \tag{7.68}$$

and from Fig. 7.9 we have

$$\frac{dw}{ds} = \sin\theta \tag{7.69}$$

Substituting (7.68) and (7.69) into (7.66) leads to

$$\frac{d^2\theta}{ds^2} = -k^2\sin\theta, \qquad k^2 = \frac{P}{EI} \tag{7.70}$$

To solve (7.70), we multiply both sides of the equation by $d\theta/ds$, i.e.,

$$\frac{d^2\theta}{ds^2}\frac{d\theta}{ds} = -k^2\sin\theta\frac{d\theta}{ds} \tag{7.71}$$

which can be written as

$$\frac{1}{2}\frac{d}{ds}\left(\frac{d\theta}{ds}\right)^2 ds = -k^2\sin\theta\,d\theta \tag{7.72}$$

Integrating (7.72) from $s = 0$ to s and the corresponding range $\theta = \alpha$ to θ, we obtain

$$\frac{1}{2}\left(\frac{d\theta}{ds}\right)^2 = -k^2\int_\alpha^\theta \sin\theta\,d\theta = k^2(\cos\theta - \cos\alpha) \tag{7.73}$$

Thus,

$$\frac{d\theta}{ds} = \pm k\sqrt{2}\sqrt{\cos\theta - \cos\alpha} \tag{7.74}$$

From Fig. 7.10, we note that $d\theta/ds$ is negative for the entire bar. Thus, the negative sign is taken.

From (7.74) we have

$$ds = -\frac{d\theta}{k\sqrt{2}\sqrt{\cos\theta - \cos\alpha}} \tag{7.75}$$

Since the deflection is symmetrical about the midspan at which $\theta = 0$, we can integrate over half of the bar and obtain

$$\frac{L}{2} = -\int_{\alpha}^{0} \frac{d\theta}{k\sqrt{2}\sqrt{\cos\theta - \cos\alpha}}$$

$$= \frac{1}{\sqrt{2}k} \int_{0}^{\alpha} \frac{d\theta}{\sqrt{\cos\theta - \cos\alpha}}$$

$$= \frac{1}{2k} \int_{0}^{\alpha} \frac{d\theta}{\sqrt{\sin^2(\alpha/2) - \sin^2(\theta/2)}} \tag{7.76}$$

in which $\cos\theta = 1 - 2\sin^2(\theta/2)$ has been used.

Define the new variable ϕ as

$$\sin\phi = \frac{1}{\beta}\sin\frac{\theta}{2} \tag{7.77}$$

where

$$\beta = \sin\frac{\alpha}{2}$$

From the relation (7.77) it is seen that when θ varies from 0 to α, the quantity $\sin\phi$ varies from 0 to 1; hence ϕ varies from 0 to $\pi/2$. Differentiate (7.77) to obtain

$$\cos\phi \, d\phi = \frac{1}{2\beta}\cos\frac{\theta}{2} \, d\theta = \frac{1}{2\beta}\sqrt{1 - \sin^2\frac{\theta}{2}} \, d\theta$$

$$= \frac{1}{2\beta}\sqrt{1 - \beta^2\sin^2\phi} \, d\theta$$

Thus,

$$d\theta = \frac{2\beta\cos\phi \, d\phi}{\sqrt{1 - \beta^2\sin^2\phi}} \tag{7.78}$$

Using (7.78) together with (7.77), the integral (7.76) takes the form

$$\frac{L}{2} = \frac{1}{k}\int_{0}^{\pi/2} \frac{d\phi}{\sqrt{1 - \beta^2\sin^2\phi}} = \frac{1}{k}F(\beta) \tag{7.79}$$

The integral $F(\beta)$ is the elliptic integral of the first kind. For a given value of β, the value of F can be found in many mathematical tables.

TABLE 7.1   Numerical solutions for P/P_{cr} and $w(L/2)/L$

α	0°	10°	20°	30°	40°	60°	80°	100°	120°	140°
P/P_{cr}	1	1.004	1.015	1.035	1.064	1.152	1.293	1.518	1.884	2.541
$w(L/2)/L$	0	0.055	0.110	0.162	0.211	0.297	0.360	0.396	0.402	0.375

Recalling the definition of k and the buckling load P_{cr} for the pinned–pinned bar, we rewrite (7.79) in the form

$$\frac{\pi}{2}\sqrt{\frac{P}{P_{cr}}} = F(\beta) \tag{7.80a}$$

or

$$\frac{P}{P_{cr}} = \frac{4}{\pi^2}F^2(\beta) \tag{7.80b}$$

The numerical solution for P/P_{cr} is given in Table 7.1 for a range of value α.
From (7.69) and (7.75), we have

$$dw = \sin\theta \, ds = -\frac{\sin\theta \, d\theta}{k\sqrt{2}\sqrt{\cos\theta - \cos\alpha}} \tag{7.81}$$

The deflection at the midspan $w(L/2)$ is

$$
\begin{aligned}
w(L/2) &= \int dw \\
&= -\frac{1}{\sqrt{2}k}\int_\alpha^0 \frac{\sin\theta \, d\theta}{\sqrt{\cos\theta - \cos\alpha}} \\
&= \frac{1}{2k}\int_0^\alpha \frac{\sin\theta \, d\theta}{\sqrt{\sin^2(\alpha/2) - \sin^2(\theta/2)}} \\
&= \frac{1}{2k}\int_0^\alpha \frac{\sin\theta \, d\theta}{\beta\sqrt{1 - \sin^2\phi}}
\end{aligned}
\tag{7.82}
$$

From (7.77) we have

$$\sin\frac{\theta}{2} = \beta\sin\phi$$

Thus,

$$\cos\frac{\theta}{2} = \sqrt{1 - \sin^2\frac{\theta}{2}} = \sqrt{1 - \beta^2\sin^2\phi}$$

and

$$\sin \theta = 2 \sin \frac{\theta}{2} \cos \frac{\theta}{2}$$

$$= 2\beta \sin \phi \sqrt{1 - \beta^2 \sin^2 \phi}$$

Using the relation above together with (7.78) in (7.82), we obtain

$$w\frac{L}{2} = \frac{2\beta}{k} \int_0^{\pi/2} \sin \phi \, d\phi = \frac{2\beta}{k}$$

$$= \frac{2 \sin(\alpha/2)}{\sqrt{P/EI}} \tag{7.83a}$$

Normalizing P with the Euler buckling load $P_{cr} = \pi^2 EI/L^2$, we have

$$\frac{w(L/2)}{L} = \frac{2 \sin(\alpha/2)}{\pi \sqrt{P/P_{cr}}} \tag{7.83b}$$

For a given value of α, the corresponding value of P/P_{cr} is obtained first using (7.80), and then from (7.83b) the midspan deflection is obtained. Some numerical solutions are given in Table 7.1. The load–deflection curve is plotted in Fig. 7.11 using the values in Table 7.1.

This load–deflection curve exhibits the postbuckling behavior of the pinned-pinned bar using the large deflection theory. It indicates that the compressive load can be increased beyond the buckling load. However, the load stays basically at the buckling load level until very large deflections (corresponding to large α angles) are produced.

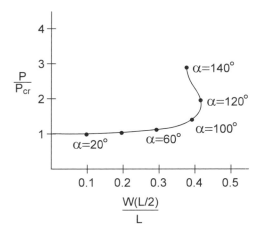

Figure 7.11 Postbuckling load–deflection curve.

In general, buckling is considered a form of structural failure, and design loads usually do not exceed buckling loads. In aircraft structures which consist of structural elements with different buckling strengths, it is not unusual to allow part of the thin or slender elements to undergo elastic postbuckling without incurring permanent deformation. Such practice would enhance the allowable compressive load for the structure.

Example 7.1   A structure consisting of four aluminum square bars rigidly connected to two heavy walls, as shown in Fig. 7.12, is designed to carry compressive load F. The cross-sections for bars 1 and 4 are 2 cm × 2 cm and for bars 2 and 3, 1 cm × 1 cm. The moments of inertia of the bars are readily obtained as

$$I_1 = I_4 = \frac{0.02 \times 0.02^3}{12} = 1.33 \times 10^{-8} \text{ m}^4$$

$$I_2 = I_3 = \frac{0.01 \times 0.01^3}{12} = 8.3 \times 10^{-10} \text{ m}^4$$

The buckling loads for the bars are

$$P_{cr1} = P_{cr4} = \frac{4\pi^2 E I_1}{L^2} = \frac{4\pi^2 \times 70 \times 10^9 \times 1.33 \times 10^{-8}}{4} = 9189 \text{ N}$$

$$P_{cr2} = P_{cr3} = \frac{4\pi^2 \times 70 \times 10^9 \times 8.3 \times 10^{-10}}{4} = 573 \text{ N}$$

Before buckling occurs, the compressive strains in the bars are equal and their loads are proportional to their respective cross-sectional areas. Thus, the loads carried by bars 1 and 4 when bars 2 and 3 have just buckled are

$$P_1 = P_4 = \frac{A_1}{A_2} P_{cr2} = \frac{0.02^2}{0.01^2} \times 573 = 2292 \text{ N}$$

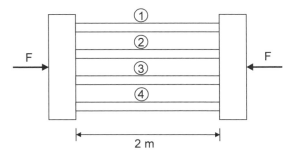

Figure 7.12   Compression of four aluminum square bars rigidly connected to walls.

At this stage the total load F is

$$F = P_1 + P_4 + P_{cr2} + P_{cr3} = 5730 \text{ N}$$

If postbuckling in bars 2 and 3 is allowed, then the load is increased to the level at which bars 1 and 4 would buckle. Recall that in the postbuckled state, the loads carried by bars 2 and 3 are basically equal to P_{cr2} and P_{cr3}, respectively. Thus, the total load capacity of this structure is

$$F = P_{cr1} + P_{cr2} + P_{cr3} + P_{cr4} = 19542 \text{ N}$$

which is more than three times the total load if no postbuckling is allowed.

7.5   BAR OF UNSYMMETRIC SECTION

Consider a straight bar of unsymmetric section under compression. The assumed perturbed deflection consists of displacement components $v(x)$ in the y-direction and $w(x)$ in the z-direction.

Take a free body of a small bar element similar to that of Fig. 7.2. The consideration of balance of forces in the y and z directions yields

$$\frac{dV_y}{dx} = 0 \tag{7.84a}$$

and

$$\frac{dV_z}{dx} = 0 \tag{7.84b}$$

respectively. The balance of moments yields

$$V_y = \frac{dM_z}{dx} - P\frac{dv}{dx} \tag{7.85a}$$

$$V_z = \frac{dM_y}{dx} - P\frac{dw}{dx} \tag{7.85b}$$

Substituting the following relations [see (4.26) and (4.27)]

$$M_y = -EI_{yz}\frac{d^2v}{dx^2} - EI_y\frac{d^2w}{dx^2}$$

$$M_z = -EI_z\frac{d^2v}{dx^2} - EI_{yz}\frac{d^2w}{dx^2}$$

into (7.85) and then (7.84), we obtain the equilibrium equations as

$$EI_z \frac{d^4v}{dx^4} + EI_{yz} \frac{d^4w}{dx^4} + P \frac{d^2v}{dx^2} = 0 \tag{7.86a}$$

$$EI_y \frac{d^4w}{dx^4} + EI_{yz} \frac{d^4v}{dx^4} + P \frac{d^2w}{dx^2} = 0 \tag{7.86b}$$

For illustration, we consider a pinned–pinned bar. It is easy to show that the displacements

$$v = C_1 \sin \frac{\pi x}{L} \tag{7.87a}$$

$$w = C_2 \sin \frac{\pi x}{L} \tag{7.87b}$$

satisfy the boundary conditions. Thus, they represent a possible buckling mode. Substitution of (7.87) into (7.86) yields

$$C_1 \left(EI_z \frac{\pi^4}{L^4} - P \frac{\pi^2}{L^2} \right) + C_2 EI_{yz} \frac{\pi^4}{L^4} = 0 \tag{7.88a}$$

$$C_1 EI_{yz} \frac{\pi^4}{L^4} + C_2 \left(EI_y \frac{\pi^4}{L^4} - P \frac{\pi^2}{L^2} \right) = 0 \tag{7.88b}$$

Eliminating C_2 from these equations, we obtain

$$C_1 \left[\left(EI_z \frac{\pi^2}{L^2} - P \right) - \frac{E^2 I_{yz}^2}{EI_y(\pi^2/L^2) - P} \times \frac{\pi^4}{L^4} \right] = 0$$

For a nontrivial solution, $C_1 \neq 0$, and we have

$$\left(EI_z \frac{\pi^2}{L^2} - P \right) \left(EI_y \frac{\pi^2}{L^2} - P \right) - E^2 I_{yz}^2 \frac{\pi^4}{L^4} = 0$$

which can be rewritten as

$$P^2 - \frac{\pi^2 E(I_y + I_z)}{L^2} P + \frac{\pi^4 E^2}{L^4}(I_y I_z - I_{yz}^2) = 0 \tag{7.89}$$

Two solutions for P are obtained:

$$P = \frac{\pi^2 E}{L^2} \left[\frac{1}{2}(I_y + I_z) \pm \frac{1}{2}\sqrt{(I_y + I_z)^2 - 4(I_y I_z - I_{yz}^2)} \right]$$

The buckling load is the smaller of the two solutions, i.e.,

$$P_{cr} = \frac{\pi^2 EI_p}{L^2} \tag{7.90}$$

where

$$I_p = \frac{1}{2}(I_y + I_z) - \frac{1}{2}\sqrt{\left(I_y - I_z\right)^2 + 4I_{yz}^2} \qquad (7.91)$$

The quantity I_p is readily recognized as the smaller moment of inertia about the principal axes of the cross-sectional area. In fact, I_p is the minimum value of the moment of inertia of the cross-section about any axis passing through the centroid. This indicates that buckling deflection takes place in the direction perpendicular to the principal axis about which the moment of inertia is the minimum.

7.6 TORSIONAL–FLEXURAL BUCKLING OF THIN-WALLED BARS

In the previous sections, we considered only lateral deflection (flexure) for possible buckling modes. For bars with low torsional rigidities, buckling may also occur by twisting, or by combined twisting and bending. This type of buckling failure is very important to aircraft structures because they often consist of thin-walled bars with open sections.

A full treatment of torsional–flexural buckling is beyond the scope of this book. In this section, we will adopt governing equations without detailed derivations and use them to perform some buckling analyses of thin-walled structures. For additional details, the reader is referred to the book by S. P. Timoshenko and J. M. Gere, *Theory of Elastic Stability*, 2nd edition, McGraw-Hill, New York, 1961.

7.6.1 Nonuniform Torsion

In Chapter 3, we discussed pure torsion, i.e., the torque is applied at the end of a bar, the ends of which are free to warp. The torque T is related to the twist angle per unit length θ as

$$T = GJ\theta = GJ\frac{d\alpha}{dx} \qquad (7.92)$$

where G is the shear modulus, J is the torsion constant, and α is the total twist angle. Since θ (and thus T also) is constant, we have the differential equation

$$\frac{d^2\alpha}{dx^2} = 0 \qquad (7.93)$$

which implies that the twist angle α is a linear function of x.

In pure torsion, only shearing stresses on cross-sections are produced, and the torque is the resultant couple of these shearing stresses. If the cross-sections are not free to rotate (e.g., a built-in end), then warping would be completely or partially suppressed, resulting in normal axial stresses (σ_{xx}) distributed over the cross-section. Additional shearing stresses are induced by the normal stresses in order to satisfy equilibrium equations. These shearing stresses produce an additional torque T' which can be related to the twist angle α as

$$T' = -EC_w\frac{d^3\alpha}{dx^3} \tag{7.94}$$

where C_w is called the **warping constant**. Thus, the total torque is

$$T = GJ\frac{d\alpha}{dx} - EC_w\frac{d^3\alpha}{dx^3} \tag{7.95}$$

If no external torques are applied between the two ends, then T is constant, and

$$\frac{dT}{dx} = GJ\frac{d^2\alpha}{dx^2} - EC_w\frac{d^4\alpha}{dx^4} = 0 \tag{7.96}$$

The differential equation above is solved in conjunction with prescribed end conditions to obtain the torsional deformation (the twist angle α). Two of the most commonly encountered end conditions are stated as follows:

$$\text{Built-in end:} \qquad \alpha = \frac{d\alpha}{dx} = 0 \tag{7.97}$$

$$\text{Simply supported end:} \qquad \alpha = \frac{d^2\alpha}{dx^2} = 0 \tag{7.98}$$

Equation (7.98) implies that the end is not allowed to rotate about the x-axis but is free to warp.

The warping constant C_w depends on the geometry of the cross-section. For cross-sections consisting of thin rectangular elements which intersect at a common point (see Fig. 7.13), the warping constant C_w is very small and

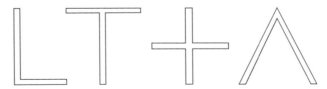

Figure 7.13   Cross-sections of thin rectangular elements with a single intersection.

can practically be taken equal to zero. Additional values of C_w for other thin-walled sections are given by Timoshenko and Gere, which are reproduced in Table 7.2.

7.6.2 Torsional Buckling of Doubly Symmetric Section

Consider a straight bar under a pair of compressive forces P applied at both ends. This is the initial state from which buckling will be examined. The method follows that of Section 7.2, i.e., a possible buckled shape (given by flexural deflections w and v, and twisting α) is assumed and then examined with the equilibrium equations and end conditions. If the buckled shape satisfies the equilibrium equations and end conditions for a given compressive force P, then buckling would occur.

If the cross-section of a bar is symmetric about the y and z axes, then its centroid and shear center coincide. An example is the bar of a cruciform cross-section with four identical flanges as shown in Fig. 7.14.

For bars of doubly symmetric cross-sections, the equilibrium equations for the buckled shape are

$$EI_z \frac{d^4v}{dx^4} + P \frac{d^2v}{dx^2} = 0 \tag{7.99}$$

$$EI_y \frac{d^4w}{dx^4} + P \frac{d^2w}{dx^2} = 0 \tag{7.100}$$

$$EC_w \frac{d^4\alpha}{dx^4} + \left(\frac{I_0}{A}P - GJ\right) \frac{d^2\alpha}{dx^2} = 0 \tag{7.101}$$

where A is the area of the cross-section, and I_0 is the polar moment of inertia about the shear center. If the shear center coincides with the centroid of the cross-section, then $I_0 = I_y + I_z$. It is evident that the three equations above are not coupled, the first two equations are equilibrium equations for flexural (bending) buckling in the y and z directions, respectively, and the torsional buckling governed by (7.101) is independent of flexural buckling. These three differential equations can be solved independently to obtain their respective critical compressive loads. The lowest value among the three critical loads is of interest in practical applications.

The equilibrium equation (7.101) for torsional buckling is of the same form as that for flexural buckling. Defining

$$p^2 = \frac{(I_0/A)P - GJ}{EC_w} \tag{7.102}$$

TABLE 7.2   Sectional properties for thin-walled sections

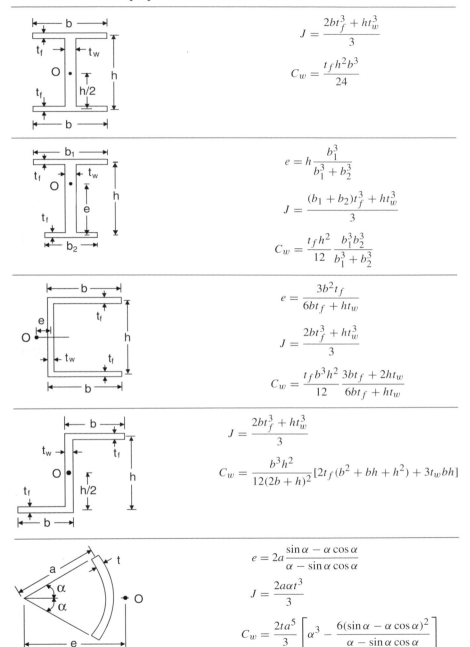

$$J = \frac{2bt_f^3 + ht_w^3}{3}$$

$$C_w = \frac{t_f h^2 b^3}{24}$$

$$e = h\frac{b_1^3}{b_1^3 + b_2^3}$$

$$J = \frac{(b_1 + b_2)t_f^3 + ht_w^3}{3}$$

$$C_w = \frac{t_f h^2}{12}\frac{b_1^3 b_2^3}{b_1^3 + b_2^3}$$

$$e = \frac{3b^2 t_f}{6bt_f + ht_w}$$

$$J = \frac{2bt_f^3 + ht_w^3}{3}$$

$$C_w = \frac{t_f b^3 h^2}{12}\frac{3bt_f + 2ht_w}{6bt_f + ht_w}$$

$$J = \frac{2bt_f^3 + ht_w^3}{3}$$

$$C_w = \frac{b^3 h^2}{12(2b + h)^2}[2t_f(b^2 + bh + h^2) + 3t_w bh]$$

$$e = 2a\frac{\sin\alpha - \alpha\cos\alpha}{\alpha - \sin\alpha\cos\alpha}$$

$$J = \frac{2a\alpha t^3}{3}$$

$$C_w = \frac{2ta^5}{3}\left[\alpha^3 - \frac{6(\sin\alpha - \alpha\cos\alpha)^2}{\alpha - \sin\alpha\cos\alpha}\right]$$

[a] O is shear center.

Source: S. P. Timoshenko and J. M. Gere, *Theory of Elastic Stability*, 2nd ed., McGraw-Hill, New York, 1961.

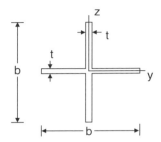

Figure 7.14 Bar of a cruciform cross-section with four identical flanges.

we can rewrite (7.101) in the form

$$\frac{d^4\alpha}{dx^4} + p^2\frac{d^2\alpha}{dx^2} = 0 \tag{7.103}$$

which is, in form, identical to (7.15). Thus, for a simply supported bar, the critical load for torsional buckling satisfies [see (7.22)]

$$pL = n\pi, \qquad n = 1, 2, 3, \ldots$$

The corresponding torsional buckling load (for $n = 1$) is given by

$$P_{cr}^T = \frac{A}{I_0}\left(GJ + \frac{\pi^2}{L^2}EC_w\right) \tag{7.104}$$

Similarly, for a bar with built-in ends, the torsional buckling load is

$$P_{cr}^T = \frac{A}{I_0}\left(GJ + \frac{4\pi^2}{L^2}EC_w\right) \tag{7.105}$$

Using the concept of effective length of buckling, L_e, torsional buckling loads for different end conditions can be expressed in the form

$$P_{cr}^T = \frac{1}{R^2}\left(GJ + \frac{\pi^2}{L_e^2}EC_w\right) \tag{7.106}$$

where

$$R = \sqrt{\frac{I_0}{A}} \tag{7.107}$$

is the **polar radius of gyration** of the cross-section.

Example 7.2 Buckling of a thin-walled bar with a cruciform cross-section as shown in Fig. 7.14 is studied. Assume that $b \gg t$. The following properties are easily obtained:

$$A = 2bt$$

$$I_z = I_y = \frac{tb^3}{12}$$

$$J = \frac{2}{3}bt^3$$

$$I_0 = I_y + I_z = \frac{tb^3}{6}$$

$$R^2 = \frac{I_0}{A} = \frac{b^2}{12}$$

From the discussion in Section 7.6.1, we have $C_w = 0$ for this cross-section if $b \gg t$. For simply supported ends, the torsional buckling load is obtained from (7.104):

$$P_{cr}^{T} = \frac{GJ}{R^2} = \frac{8t^3G}{b} \tag{a}$$

The torsional buckling load is noted to be independent of the length of the bar.

Since the bending rigidities of the cross-section of the bar are identical about both y and z axes, the flexural buckling load for the simply supported bar is obtained from (7.26) by recognizing $I_y = I_z = I$. We have

$$P_{cr} = \frac{\pi^2 EI}{L^2} = \frac{\pi^2 Etb^3}{12L^2} \tag{b}$$

For the torsional buckling to occur preceding the flexural buckling, we require

$$\frac{\pi^2 Etb^3}{12L^2} \geq \frac{8t^3G}{b}$$

or

$$\frac{L}{b} \leq \frac{\pi b}{4t}\sqrt{\frac{E}{6G}}$$

The result above indicates that torsional buckling occurs in stubby bars.

7.6.3 Torsional–Flexural Buckling

Buckling of bars with arbitrary thin-walled cross-sections usually involves coupled torsion and bending. For an arbitrary cross-section, the centroid and the shear center usually do not coincide (see Fig. 7.15).

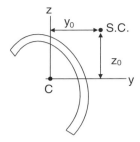

Figure 7.15   Arbitrary thin-walled cross-section.

Set up the coordinate system so that the y and z axes are the principal centroidal axes of the cross-section for which $I_{xy} = 0$. The location of the shear center is denoted by (y_0, z_0). The equilibrium equations for the buckled deflections and twist are

$$EI_z \frac{d^4v}{dx^4} + P\left(\frac{d^2v}{dx^2} + z_0\frac{d^2\alpha}{dx^2}\right) = 0 \qquad (7.108)$$

$$EI_y \frac{d^4w}{dx^4} + P\left(\frac{d^2w}{dx^2} - y_0\frac{d^2\alpha}{dx^2}\right) = 0 \qquad (7.109)$$

$$EC_w \frac{d^4\alpha}{dx^4} + \left(P\frac{I_0}{A} - GJ\right)\frac{d^2\alpha}{dx^2} + P\left(z_0\frac{d^2v}{dx^2} - y_0\frac{d^2w}{dx^2}\right) = 0 \qquad (7.110)$$

It should be noted that the polar moment of inertia I_0 is about the axis passing through the shear center. Thus,

$$I_0 = I_y + I_z + A\left(y_0^2 + z_0^2\right)$$

The three differential equations (7.108–7.110) are solved simultaneously in conjunction with the specified end conditions. For simply supported ends, we have

$$v = w = \alpha = 0 \qquad (7.111)$$

$$\frac{d^2v}{dx^2} = \frac{d^2w}{dx^2} = \frac{d^2\alpha}{dx^2} = 0 \qquad (7.112)$$

and for built-in ends we have

$$v = w = \alpha = 0 \qquad (7.113)$$

$$\frac{dv}{dx} = \frac{dw}{dx} = \frac{d\alpha}{dx} = 0 \qquad (7.114)$$

Consider the case of a bar of length L with simply supported ends. It can readily be verified that the following buckling shape functions

$$v = C_1 \sin \frac{\pi x}{L}, \qquad w = C_2 \sin \frac{\pi x}{L}, \qquad \alpha = C_3 \sin \frac{\pi x}{L} \qquad (7.115)$$

satisfy the end conditions (7.111) and (7.112). Substituting the buckling shape functions of (7.115) into (7.108)–(7.110) yields

$$\left(P - E I_z \frac{\pi^2}{L^2} \right) C_1 + P z_0 C_3 = 0 \qquad (7.116)$$

$$\left(P - E I_y \frac{\pi^2}{L^2} \right) C_2 - P y_0 C_3 = 0 \qquad (7.117)$$

$$P z_0 C_1 - P y_0 C_2 - \left(E C_w \frac{\pi^2}{L^2} + G J - \frac{I_0}{A} P \right) C_3 = 0 \qquad (7.118)$$

The equations above can be expressed in the form

$$(P - P_z) C_1 + P z_0 C_3 = 0 \qquad (7.119)$$

$$\left(P - P_y \right) C_2 - P y_0 C_3 = 0 \qquad (7.120)$$

$$P z_0 C_1 - P y_0 C_2 + \frac{I_0}{A} (P - P_\alpha) C_3 = 0 \qquad (7.121)$$

where

$$P_y = \frac{\pi^2 E I_y}{L^2}, \qquad P_z = \frac{\pi^2 E I_z}{L^2},$$

$$P_\alpha = \frac{A}{I_0} \left(G J + E C_w \frac{\pi^2}{L^2} \right) \qquad (7.122)$$

It is noted that P_y and P_z are the Euler critical loads for buckling about the y and z axes, respectively, and P_α is the critical load for pure torsional buckling.

Equations (7.119)–(7.121) are homogeneous, and a nontrivial solution for C_1, C_2, and C_3 exists only if the determinant of the coefficient matrix vanishes, i.e.,

$$\begin{vmatrix} P - P_z & 0 & P z_0 \\ 0 & P - P_y & -P y_0 \\ P z_0 & -P y_0 & I_0 (P - P_\alpha) / A \end{vmatrix} = 0 \qquad (7.123)$$

Expanding the determinant of (7.123), we obtain a cubic equation for P, which yields three possible roots. These are the critical loads of buckling. The lowest value is of practical interest.

Example 7.3   An aluminum bar of length $L = 1$ m is subjected to axial compression. The cross-section of the bar is shown in Fig. 7.16. The ends of the bar are simply supported. The material properties are given by $E = 70$ GPa and $G = 30$ GPa; the dimensions are $t = 2$ mm, $h = 0.05$ m and $b = 0.1$ m.

Since the cross-section is symmetrical about the y-axis, the centroid must lie on the y-axis. The horizontal distance $\bar{y}$ of the centroid from the vertical wall is determined by taking moments of the cross-sectional area about the vertical wall, i.e.,

$$A\bar{y} = 2\,(0.002 \times 0.1 \times 0.05) = 2 \times 10^{-5}$$

where

$$A = 3 \times 0.002 \times 0.1 = 6 \times 10^{-4}\ \text{m}^2$$

is the total cross-sectional area. Thus,

$$\bar{y} = 0.0333\ \text{m}$$

From the result of Example 5.5, we obtain the horizontal position, y_0, of the shear center as

$$y_0 = -\left(\frac{tb^2h^2}{I_y} + \bar{y}\right) = -0.076\ \text{m}$$

in which I_y is

$$I_y = \frac{t\,(2h)^3}{12} + 2tbh^2 = 1.17 \times 10^{-6}\ \text{m}^4$$

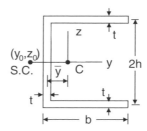

Figure 7.16   Cross-section of a bar subjected to axial compression.

Other sectional properties are

$$I_z = 2th\bar{y}^2 + 2\left(\frac{tb^3}{12}\right) + 2tb(0.05 - \bar{y})^2 = 0.67 \times 10^{-6} \text{ m}^4$$

$$I_0 = I_y + I_z + A(y_0^2 + z_0^2) = 5.32 \times 10^{-6} \text{ m}^4$$

$$J = \frac{t^3(b + 2h + b)}{3} = 8 \times 10^{-10} \text{ m}^4$$

The warping constant C_w is found from Table 7.2. We have

$$C_w = 11.9 \times 10^{-10} \text{ m}^6$$

From (7.122), we have

$$P_y = 8.08 \times 10^5 \text{ N}, \quad P_z = 4.63 \times 10^5 \text{ N}, \quad P_\alpha = 9.54 \times 10^4 \text{ N} \qquad \text{(a)}$$

Noting that $y_0 = -0.076$ m and $z_0 = 0$, we see that the three equilibrium equations, (7.119) to (7.121), become

$$(P - P_z)C_1 = 0 \qquad \text{(b)}$$

$$(P - P_y)C_2 + 0.076PC_3 = 0 \qquad \text{(c)}$$

$$0.076PC_2 + 0.0089(P - P_\alpha)C_3 = 0 \qquad \text{(d)}$$

It is evident that (b) is not coupled with the other two equations. The solution for (b) is

$$P_{cr} = P_z = 4.63 \times 10^5 \text{ N} \qquad \text{(e)}$$

In order to have a nontrivial solution for (c) and (d), we require that

$$\begin{vmatrix} P - P_y & 0.076\,P \\ 0.076\,P & 0.0089\,(P - P_\alpha) \end{vmatrix} = 0$$

After expanding the determinant, we obtain

$$P^2 - 25.74 \times 10^5 P + 21.97 \times 10^{10} = 0 \qquad \text{(f)}$$

The two possible roots for P are

$$P_{cr} = 0.89 \times 10^5 \text{ N} \qquad \text{(g)}$$

and

$$P_{cr} = 24.8 \times 10^5 \text{ N} \qquad \text{(h)}$$

Among the three critical loads, the lowest value is $P_{cr} = 0.89 \times 10^5$ N, which is lower than any of the uncoupled buckling loads given by (a). Thus, the buckling load of the bar is reduced by the coupling of torsion and bending.

7.7   ELASTIC BUCKLING OF FLAT PLATES

Thin panels under compression would also buckle. Since aircraft structures are composed of thin-walled components, buckling of thin sheets is an important subject of study to aeronautical engineers. Unlike buckling of slender bars, buckling analysis of thin sheets requires knowledge of advanced structural mechanics such as theories of plates and shells as well as advanced mathematics. These subjects are usually covered in graduate-level courses and are thus beyond the scope of this book.

The objective of this section is to provide the student with some exposure to the subject of buckling of thin plates. To achieve this without requiring an advanced background of the student, many derivations of the equations and formulas will be skipped. Whenever possible, the column buckling results discussed in previous sections will be used to provide some qualitative explanations.

7.7.1   Governing Equation for Flat Plates

Consider a flat plate of thickness h under in-plane line loads (force/length) as shown in Fig. 7.17. This is considered the initial state of the flat plate. If additional deformation is produced in terms of transverse deflection w, then transverse shear forces Q_x and Q_y (force/length), bending moments M_x and M_y, and twisting moment M_{xy} (moment/length) are induced in the plate. The positive directions of these plate resultant forces and moments are shown in Fig. 7.18. These induced plate resultant shear forces and moments together with the existing in-plane resultant forces N_x, N_y, and N_{xy} (force/length) must satisfy the equilibrium equations. Since the initial state under in-plane

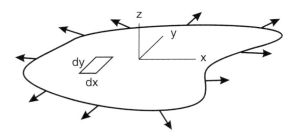

Figure 7.17   Flat plate under in-plane loads.

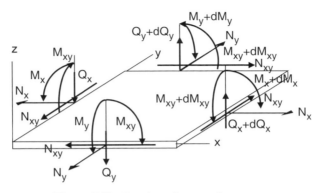

Figure 7.18 Resultant forces and moments.

forces is already in a state of equilibrium, the only equilibrium equation to be satisfied is the balance of forces in the z-direction. Following a procedure similar to that used in Section 7.2 for straight bars, we obtain the equilibrium equation for the differential plate element of Fig. 7.18 as

$$\frac{\partial^4 w}{\partial x^4} + 2\frac{\partial^4 w}{\partial x^2 \partial y^2} + \frac{\partial^4 w}{\partial y^4} = \frac{1}{D}\left(N_x \frac{\partial^2 w}{\partial x^2} + N_y \frac{\partial^2 w}{\partial y^2} + 2N_{xy}\frac{\partial^2 w}{\partial x \partial y}\right) \quad (7.124)$$

where

$$D = \frac{Eh^3}{12(1 - v^2)} \quad (7.125)$$

is the **bending rigidity** of the plate.

Boundary Conditions   The general solution to the differential equation (7.124) contains arbitrary constants which are to be determined from the boundary conditions of the plate. For illustration, we consider the boundary conditions along the edge $x = 0$.

Clamped Edge   A clamped edge means that the deflection (w) and rotation $(\partial w/\partial x)$ along this edge are not allowed during plate deformation. Thus, the boundary conditions are given by

$$(w)_{x=0} = 0, \qquad \left(\frac{\partial w}{\partial x}\right)_{x=0} = 0 \quad (7.126)$$

Simply Supported Edge   The deflection along a simply supported edge is zero, but the edge is allowed to rotate freely about the y-axis; i.e., there are no bending moments M_x along this edge. These conditions are expressed in terms of plate displacement in the form

$$(w)_{x=0} = 0, \qquad M_x = -D\left(\frac{\partial^2 w}{\partial x^2} + v\frac{\partial^2 w}{\partial y^2}\right)_{x=0} = 0 \qquad (7.127)$$

Free Edge   If the edge $x = 0$ is free from external loads, then the shear force and bending moment must vanish along this edge. Since $\partial M_{xy}/\partial y$ produces an equivalent shear force action, the total shear force acting on this edge must include Q_x and $\partial M_{xy}/\partial y$. Thus, the conditions for a free edge are

$$M_x = -D\left(\frac{\partial^2 w}{\partial x^2} + v\frac{\partial^2 w}{\partial y^2}\right)_{x=0} = 0 \qquad (7.128)$$

$$\left(Q_x - \frac{\partial M_{xy}}{\partial y}\right)_{x=0} = 0 \qquad (7.129)$$

By using the relations

$$M_{xy} = D(1 - v)\frac{\partial^2 w}{\partial x\, \partial y}$$

and

$$Q_x = -D\frac{\partial}{\partial x}\left(\frac{\partial^2 w}{\partial x^2} + \frac{\partial^2 w}{\partial y^2}\right)$$

the condition (7.129) can be written as

$$\left[\frac{\partial^3 w}{\partial x^3} + (2 - v)\frac{\partial^3 w}{\partial x\, \partial y^2}\right]_{x=0} = 0 \qquad (7.130)$$

7.7.2   Cylindrical Bending

In general, buckling of flat plates is a 2-D problem. The deflection surface of a buckled plate is a function of both x and y coordinates. A special case of a rectangular plate under uniform compression $N_x^0 = -N_x$ as shown in Fig. 7.19 may be treated approximately as a 1-D problem if $b \gg a$. For such a case, the deflection w can be assumed to be independent of the y-axis. The deformed plate forms a cylindrical surface about the y-axis and thus the name **cylindrical bending**. Such an assumption is good except for the regions near the edges $y = 0$ and $y = b$. Neglecting this edge effect and assuming that $w(x)$ is a function of x only, the governing equation (7.124) reduces to

$$D\frac{d^4 w}{dx^4} + N_x^0\frac{d^2 w}{dx^2} = 0 \qquad (7.131)$$

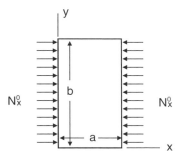

Figure 7.19   Rectangular plate under uniform compression.

Multiplying (7.131) by b, we have

$$Db\frac{d^4w}{dx^4} + P\frac{d^2w}{dx^2} = 0 \qquad (7.132)$$

where $P = N_x^0 b$ is the total compressive force. Comparing (7.132) with (7.15), we note that if the bending rigidity EI of the bar is replaced by the total bending rigidity Db of the plate, then the buckling of a plate in cylindrical bending is identical to buckling of a straight bar.

From (7.125) we note that

$$Db = \frac{Eh^3 b}{12(1 - v^2)} = \frac{EI}{1 - v^2} \qquad (7.133)$$

Thus the only difference between a bar and a plate in cylindrical bending is a factor of $1 - v^2$. This is the result of the assumption that the bending stress in the plate is in a state of plane strain parallel to the x–z plane while in the bar a state of plane stress is assumed.

7.7.3   Buckling of Rectangular Plates

Simply Supported Edges   Consider a rectangular plate compressed by uniform in-plane forces N_x^0 along the edges $x = 0$ and $x = a$ as shown in Fig. 7.20. The four edges are assumed to be *simply supported*. The deflection surface of the buckled plate is given by

$$w = C_{mn} \sin\frac{m\pi x}{a} \sin\frac{n\pi y}{b}, \qquad m, n = 1, 2, 3, \ldots \qquad (7.134)$$

It is easy to verify that this deflection function satisfies the simply supported boundary conditions (7.127) along all four edges. It must also satisfy

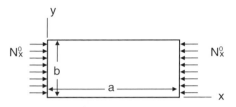

Figure 7.20 Simply supported rectangular plate compressed by uniform in-plane forces.

the equilibrium equation (7.124). Substitution of (7.134) in (7.124) leads to

$$\frac{m^4\pi^4}{a^4} + 2\frac{m^2}{a^2}\frac{n^2}{b^2}\pi^4 + \frac{n^4\pi^4}{b^4} = \frac{1}{D}N_x^0\frac{m^2\pi^2}{a^2} \qquad (7.135)$$

from which we obtain the critical value of the compressive force

$$N_x^0 = \frac{\pi^2 a^2 D}{m^2}\left(\frac{m^2}{a^2} + \frac{n^2}{b^2}\right)^2 \qquad (7.136)$$

The solution given by (7.136) represents the buckling force associated with the buckling mode shape in the form of (7.134) with a combination of m and n. From (7.136), it is obvious that among all the possible values of n, $n = 1$ will make N_x^0 the smallest. Thus the critical value of the compressive force becomes

$$N_x^0 = \frac{\pi^2 D}{b^2}\left(m\frac{b}{a} + \frac{1}{m}\frac{a}{b}\right)^2 = k\frac{\pi^2 D}{b^2} \qquad (7.137)$$

where

$$k = \left(m\frac{b}{a} + \frac{1}{m}\frac{a}{b}\right)^2 \qquad (7.138)$$

If $b \gg a$, then N_x^0 can be approximated by the expression

$$N_x^0 = \frac{\pi^2 D}{a^2}m^2 \qquad (7.139)$$

which can easily be verified as the solution to the buckling equation (7.131) for the cylindrical bending problem. In such cases, $m = 1$ yields the lowest critical force; i.e.,

$$\left(N_x^0\right)_{cr} = \frac{\pi^2 D}{a^2} \qquad (7.140)$$

For plates with general aspect ratios, the minimum critical force N_x^0 depends on the ratio a/b and the mode number m. If a is smaller than b, the second term in parentheses in (7.138) is always smaller than the first term. Then the minimum value of N_x^0 can be obtained by taking the minimum value of the first term, i.e., $m = 1$. Thus for $a \leq b$, we have

$$\left(N_x^0\right)_{cr} = \frac{\pi^2 D}{b^2} \left(\frac{b}{a} + \frac{a}{b}\right)^2 \tag{7.141}$$

whose minimum value occurs when $a = b$ for a constant b. This conclusion can easily be obtained by setting $\partial\left(N_x^0\right)_{cr} / \partial(a/b) = 0$. This result indicates that the buckling load for a plate of a given width is the smallest if the plate is square and is given by

$$\left(N_x^0\right)_{cr} = \frac{4\pi^2 D}{a^2} \tag{7.142}$$

This is four times the buckling load for plates under cylindrical bending.

To find the minimum critical values of N_x^0 for other aspect ratios a/b, we must minimize the factor k given in (7.138). Figure 7.21 shows the plot of k versus aspect ratio a/b for various mode numbers m. It is interesting to note that the buckling mode number switches from $m = 1$ to $m = 2$ at $a/b = \sqrt{2}$, and from $m = 2$ to $m = 3$ at $a/b = \sqrt{6}$, and so on. In fact, a general relation exists; i.e.,

$$\frac{a}{b} = \sqrt{m(m + 1)} \tag{7.143}$$

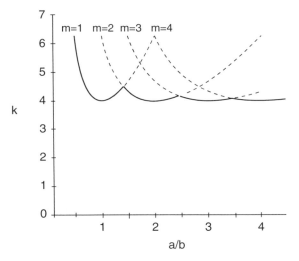

Figure 7.21 Plot of k versus aspect ratio for various mode numbers.

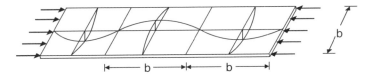

Figure 7.22   Buckled deflection surface $a \gg b$.

If $a \gg b$, m becomes large, and then

$$m \simeq \frac{a}{b} \tag{7.144}$$

The corresponding buckled deflection surface is

$$w = C \sin \frac{m\pi x}{a} \sin \frac{\pi y}{b} = C \sin \frac{\pi x}{b} \sin \frac{\pi y}{b} \tag{7.145}$$

The preceding buckled shape indicates that a very long plate buckles as if the plate were divided into many simply supported square plates of size $b \times b$. A sketch illustrating the buckled deflection surface is shown in Fig. 7.22.

Other Boundary Conditions

Simply Supported Edges along $x = 0$, $x = a$, and $y = 0$, and the Free Edge along $y = b$ We now change the edge of the plate (see Fig. 7.20) along $y = b$ to the free edge condition while keeping all other three edges simply supported. The lowest critical compressive force N_x^0 is given by the general form

$$\left(N_x^0\right)_{cr} = k \frac{\pi^2 D}{b^2} \tag{7.146}$$

The value of the buckling coefficient k depends on the aspect ratio a/b and Poisson's ratio ν. Table 7.3 lists the values of k for various aspect ratios and $\nu = 0.25$. For long plates (i.e., $a \gg b$) the value of k can be calculated using the approximate formula

$$k = 0.456 + \frac{b^2}{a^2} \tag{7.147}$$

TABLE 7.3   Values of k for simply supported edge along $y = 0$ and free edge along $y = b$, $\nu = 0.25$

a/b	0.50	1.0	1.2	1.4	1.6	1.8	2.0	2.5	3.0	4.0	5.0
k	4.40	1.440	1.135	0.952	0.835	0.755	0.698	0.610	0.564	0.516	0.506

TABLE 7.4   Values of k for clamped edge along $y = 0$ and free edge along $y = b$, $\nu = 0.25$

a/b	1.0	1.1	1.2	1.3	1.4	1.5	1.6	1.7	1.8	1.9	2.0	2.2	2.4
k	1.70	1.56	1.47	1.41	1.36	1.34	1.33	1.33	1.34	1.36	1.38	1.45	1.47

Simply Supported Edges Along $x = 0$, $x = a$, Clamped Edge $y = 0$ and the Free Edge $y = b$ For a rectangular plate with these boundary conditions, the values of k for a/b ratios up to 2.4 are listed in Table 7.4.

7.7.4 Buckling under Shearing Stresses

Consider a rectangular plate under uniform shear forces N_{xy} along all edges that are simply supported (see Fig. 7.23). The buckling is actually caused by the compressive stresses on the planes at 45° against the x-axis. The buckled deflection surface should satisfy the equilibrium equation (7.124). In this case, (7.124) becomes

$$\frac{\partial^4 w}{\partial x^4} + 2\frac{\partial^4 w}{\partial x^2 \partial y^2} + \frac{\partial^4 w}{\partial y^4} = \frac{2N_{xy}}{D}\frac{\partial^2 w}{\partial x \partial y} \qquad (7.148)$$

For this case, a single term such as that given by (7.134) cannot satisfy (7.148), although the simply supported boundary conditions are satisfied. There are approximate methods to find the solution for the buckling load, which can be given as

$$\left(N_{xy}\right)_{\text{cr}} = \pm k\frac{\pi^2 D}{b^2} \qquad (7.149)$$

where k is a constant depending on the ratio a/b. The two signs indicate that the buckling load does not depend on the direction of the shearing force. In Table 7.5, values of k for various aspect ratios are listed.

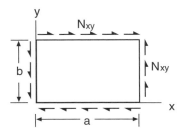

Figure 7.23   Rectangular plate under uniform shear forces.

TABLE 7.5   **Values of k for buckling under shearing forces**

a/b	1.0	1.2	1.4	1.5	1.6	1.8	2.0	2.5	3.0	4.0	∞
k	9.34	8.0	7.3	7.1	7.0	6.8	6.6	6.1	5.9	5.7	5.35

7.8   LOCAL BUCKLING OF OPEN SECTIONS

Thin flat panels are inefficient in carrying compressive loads because their buckling stresses are low. However, thin-walled sections formed with thin sheets, such as angles and channels, can provide much improved compressive buckling strengths.

The global buckling of slender bars of thin-walled sections was discussed in Section 7.6. For short open-section members, local buckling may occur before global buckling. Local buckling often takes the form of buckling of flat plates.

Let us consider an open section composed of flat sheet elements. It has been found that the buckling load of the open section can be approximated reasonably well by summing the individual buckling loads of the sheet elements. The rationale for such an approach is similar to that used in Example 7.1.

Denoting the buckling load for sheet element i by $P_{cr}^{(i)}$, the total buckling load of the composite section is estimated as

$$P_{cr} = P_{cr}^{(1)} + P_{cr}^{(2)} + P_{cr}^{(3)} + \cdots \tag{7.150}$$

The (average) **crippling stress** σ_{cr} of the section is

$$\sigma_{cr} = \frac{P_{cr}}{A} \tag{7.151}$$

where A is the total cross-sectional area.

To estimate the local buckling load, boundary conditions for each sheet element must be specified. This is usually a rather difficult task because restraints from the adjacent sheet elements, in general, do not fall into the usual category of edge conditions for flat plates. Approximate but conservative boundary conditions are usually used. For example, the angle section shown in Fig. 7.24 has two identical flanges which would buckle under the same compressive load. In this case, there is no restraint from one flange on the other. Hence, the edge along the junction can be assumed to be simply supported. For more general sections, the simply supported edge is often used to approximate the boundary conditions along the junction of two sheet elements.

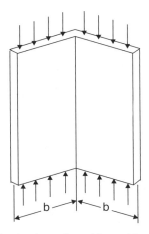

Figure 7.24 Angle section with two identical flanges.

Example 7.4 We will estimate the crippling load for a channel section shown in Fig. 7.25. The material is an aluminum alloy with $E = 69$ GPa and $\nu = 0.3$.

The channel is assumed to be the assembly of three flat plate elements. The loading edges (top and bottom) are simply supported; the edges along the junctions between elements are approximated as simply supported edges; the free edges in elements 1 and 3 are obvious. The local buckling load for each element is calculated as follows.

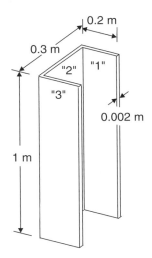

Figure 7.25 Channel section.

Elements 1 and 3 These two plate elements are simply supported along three sides and free on one side. The aspect ratio a/b is 5.0 which gives the buckling coefficient $k = 0.506$ (see Table 7.3). The bending rigidity of the plate element is given by (7.125). We have

$$D = \frac{Eh^3}{12(1-v^2)} = \frac{69 \times 10^9 \times (0.002)^3}{12(1-0.09)} = 50.5 \text{ N·m}$$

Thus, the buckling (line) load for plate elements 1 and 3 is

$$\left(N_x^0\right)_{cr} = k\frac{\pi^2 D}{b^2} = \frac{0.506 \times \pi^2 \times 50.5}{0.04} = 6305 \text{ N/m}$$

The local buckling load for both elements is

$$P_{cr}^{(1)} = P_{cr}^{(3)} = 6305 \times 0.2 = 1261 \text{ N}$$

Element 2 Plate element 2 is simply supported along the four edges. The aspect ratio is $a/b = 3.33$. The corresponding buckling coefficient k can be picked up from Fig. 7.21. We have $k \simeq 4.0$. Thus, the local buckling (line) load for element 2 is

$$\left(N_x^0\right)_{cr} = k\frac{\pi^2 D}{b^2} = \frac{4.0 \times \pi^2 \times 50.5}{0.09} = 22,155 \text{ N}$$

The buckling load for element 2 is

$$P_{cr}^{(2)} = 22,155 \times 0.3 = 6646 \text{ N}$$

Using (7.150), the total crippling load for the channel section is

$$P_{cr} = P_{cr}^{(1)} + P_{cr}^{(2)} + P_{cr}^{(3)} = 9168 \text{ N}$$

PROBLEMS

7.1 The truss structure consists of two bars connected by a pin-joint (which allows free rotation of the bars). The other ends of the bars are hinged as shown in Fig. 7.26. A weight W is hung at the joint. Find the maximum weight the truss can sustain before buckling occurs.

7.2 A bar is built-in at the left end and supported at the right end by a linear spring with spring constant α. Find the equation for buckling loads. *Hint:* The boundary conditions are $w = 0$ and $dw/dx = 0$ at the left end; and $M = 0$ and $V = -\alpha w$ at the right end.

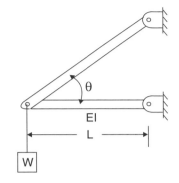

Figure 7.26 Two-bar truss.

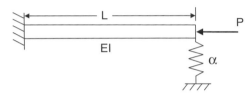

Figure 7.27 Bar with a built-in end and an elastically supported end.

7.3 Two steel bars ($E = 210$ GPa) are connected by a hinge as shown in Fig. 7.28. The square cross-section of the bar is $5\,\text{cm} \times 5\,\text{cm}$. Find the buckling load for the bar with a built-in end.

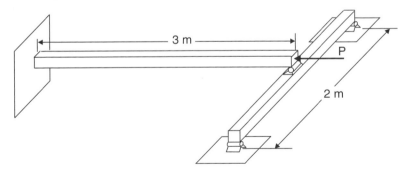

Figure 7.28 Two-bar structure.

7.4 Find the buckling load equation for the bar with the left end built-in and the right end simply supported but constrained by a rotational spring (see Fig. 7.29). The spring constant β relates the bending moment M and the rotation $\theta = dw/dx$ by $M = \beta\theta$.

7.5 Two steel bars of a 4-cm circular cross-section are rigidly connected into a T-shaped structure. The diameter of the bars is 4 cm. Three ends are

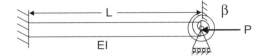

Figure 7.29 Bar with a built-in end and a rotationally constrained end.

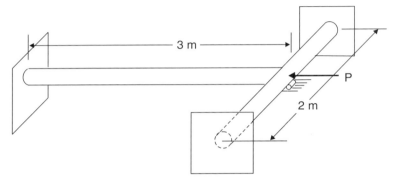

Figure 7.30 Structure with two rigidly connected bars.

built-in as shown in Fig. 7.30. At the joint, a roller support is provided to prevent vertical deflection of the joint. Compression is applied as shown in the figure. Find the lowest buckling load.

7.6 For the structure of Problem 7.5, find the buckling load if the roller support at the joint is removed.

7.7 A simply supported bar has a doubly symmetrical cross-section consisting of a thin web and thin flanges as shown in Fig. 7.31. Find the length of the bar at which the flexural buckling load is equal to the torsional buckling load.

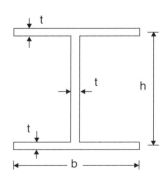

Figure 7.31 Cross-section of a simply supported bar.

7.8 A simply supported aluminum ($E = 70$ GPa, $G = 27$ GPa) bar 2 m in length has the cross-section shown in Fig. 7.32. Find the lowest three buckling loads.

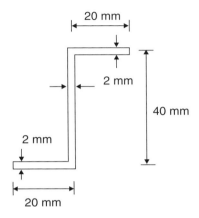

Figure 7.32   Cross-section of a thin-walled bar.

7.9 Find the buckling load of a 1-m-long bar having a thin-walled circular cross-section 50 mm in diameter and wall thickness of 2 mm. If the closed section is made into an open one by cutting a longitudinal slit over the entire length of the bar, what is the buckling load? Assume that $E = 70$ GPa and $G = 27$ GPa.

8

ANALYSIS OF COMPOSITE LAMINATES

8.1 PLANE STRESS EQUATIONS FOR COMPOSITE LAMINA

Many structural applications of fiber-reinforced composite materials are in the form of thin laminates, and a state of plane stress parallel to the laminate can be assumed with reasonable accuracy. For this reason, formulations in plane stress are of particular interest for composite structures.

In this chapter, the x_1-axis will be designated as the fiber direction as shown in Fig. 8.1. For a state of plane stress parallel to the x_1–x_2 plane in an orthotropic solid (i.e., $\sigma_{33} = \sigma_{13} = \sigma_{23} = 0$), the stress-strain relations are given by (2.109) or (2.111). Symbolically, (2.111) can be expressed in

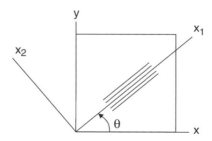

Figure 8.1 Fiber direction and coordinate systems.

the form

$$
\left\{ \begin{array}{c} \sigma_{11} \\ \sigma_{22} \\ \sigma_{12} \end{array} \right\} = \left[\begin{array}{ccc} Q_{11} & Q_{12} & 0 \\ Q_{21} & Q_{22} & 0 \\ 0 & 0 & Q_{66} \end{array} \right] \left\{ \begin{array}{c} \varepsilon_{11} \\ \varepsilon_{22} \\ \gamma_{12} \end{array} \right\} \tag{8.1}
$$

where

$$
\begin{aligned}
Q_{11} &= \frac{E_1}{1 - \nu_{12}\nu_{21}}, & Q_{12} = Q_{21} &= \frac{\nu_{12}E_2}{1 - \nu_{12}\nu_{21}} \\
Q_{22} &= \frac{E_2}{1 - \nu_{12}\nu_{21}}, & Q_{66} &= G_{12} \\
Q_{16} &= Q_{26} = Q_{61} = Q_{62} = 0
\end{aligned} \tag{8.2}
$$

are called **reduced stiffnesses**, which should not be confused with elastic constants c_{ij}.

The inverse relations of (8.1) are given by (2.109), which are usually presented as

$$
\left\{ \begin{array}{c} \varepsilon_{11} \\ \varepsilon_{22} \\ \gamma_{12} \end{array} \right\} = \left[\begin{array}{ccc} S_{11} & S_{12} & 0 \\ S_{21} & S_{22} & 0 \\ 0 & 0 & S_{66} \end{array} \right] \left\{ \begin{array}{c} \sigma_{11} \\ \sigma_{22} \\ \sigma_{12} \end{array} \right\} \tag{8.3}
$$

where

$$
\begin{aligned}
S_{11} &= \frac{1}{E_1}, & S_{12} &= -\frac{\nu_{21}}{E_2}, & S_{16} &= 0 \\
S_{21} &= -\frac{\nu_{12}}{E_1}, & S_{22} &= \frac{1}{E_2}, & S_{26} &= 0 \\
S_{61} &= 0, & S_{62} &= 0, & S_{66} &= \frac{1}{G_{12}}
\end{aligned} \tag{8.4}
$$

is a subset of the elastic compliances a_{ij} ($i, j = 1, 2, \ldots, 6$) in (2.84).

Note that in the plane stress-strain relations for orthotropic solids in a state of plane stress, there are four independent material constants. For fibrous composites, E_1 (Young's modulus in the fiber direction), E_2 (transverse Young's modulus), G_{12} (longitudinal shear modulus), and ν_{12} (transverse/longitudinal Poisson ratio) are often used to characterize the composite in a state of plane stress.

In stress analyses, sometimes a coordinate system x–y is set up which does not always coincide with the material principal axes, x_1 and x_2 as illustrated in Fig. 8.1. The two sets of stress components with respect to these two coordinates systems are related by the transformation matrix $[T_\sigma]$ [see (2.61)]:

$$\left\{ \begin{array}{c} \sigma_{11} \\ \sigma_{22} \\ \sigma_{12} \end{array} \right\} = [T_\sigma] \left\{ \begin{array}{c} \sigma_{xx} \\ \sigma_{yy} \\ \sigma_{xy} \end{array} \right\} \tag{8.5}$$

where

$$[T_\sigma] = \begin{bmatrix} \cos^2 \theta & \sin^2 \theta & 2\sin\theta\cos\theta \\ \sin^2 \theta & \cos^2 \theta & -2\sin\theta\cos\theta \\ -\sin\theta\cos\theta & \sin\theta\cos\theta & \cos^2 \theta - \sin^2 \theta \end{bmatrix} \tag{8.6}$$

In the same manner, the strains with respect to the two coordinate systems are related by

$$\left\{ \begin{array}{c} \varepsilon_{11} \\ \varepsilon_{22} \\ \gamma_{12} \end{array} \right\} = [T_\varepsilon] \left\{ \begin{array}{c} \varepsilon_{xx} \\ \varepsilon_{yy} \\ \gamma_{xy} \end{array} \right\} \tag{8.7}$$

where

$$[T_\varepsilon] = \begin{bmatrix} \cos^2 \theta & \sin^2 \theta & \sin\theta\cos\theta \\ \sin^2 \theta & \cos^2 \theta & -\sin\theta\cos\theta \\ -2\sin\theta\cos\theta & 2\sin\theta\cos\theta & \cos^2 \theta - \sin^2 \theta \end{bmatrix} \tag{8.8}$$

Note that the inverses $[T_\sigma]^{-1}$ and $[T_\varepsilon]^{-1}$ can be obtained by replacing θ in (8.6) and (8.8) with $-\theta$.

Using the transformation matrices $[T_\sigma]$ and $[T_\varepsilon]$, we have

$$\left\{ \begin{array}{c} \sigma_{xx} \\ \sigma_{yy} \\ \sigma_{xy} \end{array} \right\} = [T_\sigma]^{-1} \left\{ \begin{array}{c} \sigma_{11} \\ \sigma_{22} \\ \sigma_{12} \end{array} \right\} = [T_\sigma]^{-1}[Q] \left\{ \begin{array}{c} \varepsilon_{11} \\ \varepsilon_{22} \\ \gamma_{12} \end{array} \right\}$$

$$= [T_\sigma]^{-1}[Q][T_\varepsilon] \left\{ \begin{array}{c} \varepsilon_{xx} \\ \varepsilon_{yy} \\ \gamma_{xy} \end{array} \right\}$$

Thus, the stress-strain relations for the state of plane stress parallel to $x-y$ (x_1-x_2) plane become

$$\left\{ \begin{array}{c} \sigma_{xx} \\ \sigma_{yy} \\ \sigma_{xy} \end{array} \right\} = [\bar{Q}] \left\{ \begin{array}{c} \varepsilon_{xx} \\ \varepsilon_{yy} \\ \gamma_{xy} \end{array} \right\} \tag{8.9}$$

where

$$[\bar{Q}] = [T_\sigma]^{-1} [Q] [T_\varepsilon] \tag{8.10}$$

The explicit expressions for the elements in $[Q]$ are given by

$$\bar{Q}_{11} = Q_{11} \cos^4 \theta + 2(Q_{12} + 2Q_{66}) \sin^2 \theta \cos^2 \theta + Q_{22} \sin^4 \theta$$

$$\bar{Q}_{12} = \bar{Q}_{21} = (Q_{11} + Q_{22} - 4Q_{66}) \sin^2 \theta \cos^2 \theta + Q_{12}(\sin^4 \theta + \cos^4 \theta)$$

$$\bar{Q}_{22} = Q_{11} \sin^4 \theta + 2(Q_{12} + 2Q_{66}) \sin^2 \theta \cos^2 \theta + Q_{22} \cos^4 \theta \tag{8.11}$$

$$\bar{Q}_{16} = \bar{Q}_{61}$$

$$= (Q_{11} - Q_{12} - 2Q_{66}) \sin \theta \cos^3 \theta + (Q_{12} - Q_{22} + 2Q_{66}) \sin^3 \theta \cos \theta$$

$$\bar{Q}_{26} = \bar{Q}_{62}$$

$$= (Q_{11} - Q_{12} - 2Q_{66}) \sin^3 \theta \cos \theta + (Q_{12} - Q_{22} + 2Q_{66}) \sin \theta \cos^3 \theta$$

$$\bar{Q}_{66} = (Q_{11} + Q_{22} - 2Q_{12} - 2Q_{66}) \sin^2 \theta \cos^2 \theta + Q_{66}(\sin^4 \theta + \cos^4 \theta)$$

The fact that $[\bar{Q}]$ is a full matrix indicates that the in-plane shear deformation γ_{xy} is coupled with the normal deformations ε_{xx} and ε_{yy}. This behavior is called **shear-extension coupling**.

Following a similar procedure, we obtain

$$\left\{ \begin{array}{c} \varepsilon_{xx} \\ \varepsilon_{yy} \\ \gamma_{xy} \end{array} \right\} = [\bar{S}] \left\{ \begin{array}{c} \sigma_{xx} \\ \sigma_{yy} \\ \sigma_{xy} \end{array} \right\} \tag{8.12}$$

where

$$[\bar{S}] = [T_\varepsilon]^{-1} [S] [T_\sigma] \tag{8.13}$$

and

$$\bar{S}_{11} = S_{11} \cos^4 \theta + (2S_{12} + S_{66}) \sin^2 \theta \cos^2 \theta + S_{22} \sin^4 \theta$$

$$\bar{S}_{12} = S_{21} = S_{12}(\sin^4 \theta + \cos^4 \theta) + (S_{11} + S_{22} - S_{66}) \sin^2 \theta \cos^2 \theta$$

$$\bar{S}_{22} = S_{11} \sin^4 \theta + (2S_{12} + S_{66}) \sin^2 \theta \cos^2 \theta + S_{22} \cos^4 \theta \tag{8.14}$$

$$\bar{S}_{16} = \bar{S}_{61} = (2S_{11} - 2S_{12} - S_{66}) \sin \theta \cos^3 \theta + (2S_{12} - 2S_{22} + S_{66}) \sin^3 \theta \cos \theta$$

$$\bar{S}_{26} = \bar{S}_{62} = (2S_{11} - 2S_{12} - S_{66}) \sin^3 \theta \cos \theta + (2S_{12} - 2S_{22} + S_{66}) \sin \theta \cos^3 \theta$$

$$\bar{S}_{66} = 2(2S_{11} + 2S_{22} - 4S_{12} - S_{66}) \sin^2 \theta \cos^2 \theta + S_{66}(\sin^4 \theta + \cos^4 \theta)$$

By using the definitions of engineering moduli, the stress-strain relations in an arbitrary coordinate system (x, y) can also be expressed in the form

$$
\left\{ \begin{array}{c} \varepsilon_{xx} \\ \varepsilon_{yy} \\ \gamma_{xy} \end{array} \right\} = \begin{bmatrix} \dfrac{1}{E_x} & -\dfrac{\nu_{yx}}{E_y} & \dfrac{\eta_{xy,x}}{G_{xy}} \\[2ex] -\dfrac{\nu_{xy}}{E_x} & \dfrac{1}{E_y} & \dfrac{\eta_{xy,y}}{G_{xy}} \\[2ex] \dfrac{\eta_{x,xy}}{E_x} & \dfrac{\eta_{y,xy}}{E_y} & \dfrac{1}{G_{xy}} \end{bmatrix} \left\{ \begin{array}{c} \sigma_{xx} \\ \sigma_{yy} \\ \sigma_{xy} \end{array} \right\} \tag{8.15}
$$

Since the x- or y-axis may not coincide with the principal (fiber) direction, the engineering moduli E_x, E_y, ν_{xy}, G_{xy}, and $\eta_{xy,x}$ ... are called the **apparent engineering moduli**. Comparing (8.12) and (8.15), we have

$$
\bar{S}_{11} = \frac{1}{E_x}, \qquad \bar{S}_{12} = -\frac{\nu_{yx}}{E_y}, \qquad \bar{S}_{16} = \frac{\eta_{xy,x}}{G_{xy}}
$$

$$
\bar{S}_{21} = -\frac{\nu_{xy}}{E_x}, \qquad \bar{S}_{22} = \frac{1}{E_y}, \qquad \bar{S}_{26} = \frac{\eta_{xy,y}}{G_{xy}} \tag{8.16}
$$

$$
\bar{S}_{61} = \frac{\eta_{x,xy}}{E_x}, \qquad \bar{S}_{62} = \frac{\eta_{y,xy}}{E_y}, \qquad \bar{S}_{66} = \frac{1}{G_{xy}}
$$

For an orthotropic material, the apparent engineering moduli can be expressed in terms of the principal engineering moduli through the use of (8.16), (8.14), and (8.4). The relations are

$$
E_x = \left[\frac{1}{E_1} \cos^4 \theta + \left(\frac{1}{G_{12}} - \frac{2\nu_{12}}{E_1} \right) \sin^2 \theta \cos^2 \theta + \frac{1}{E_2} \sin^4 \theta \right]^{-1}
$$

$$
\nu_{xy} = E_x \left[\frac{\nu_{12}}{E_1} - \left(\frac{1}{E_1} + \frac{1}{E_2} + \frac{2\nu_{12}}{E_1} - \frac{1}{G_{12}} \right) \sin^2 \theta \cos^2 \theta \right]
$$

$$
E_y = \left[\frac{1}{E_1} \sin^4 \theta + \left(\frac{1}{G_{12}} - \frac{2\nu_{12}}{E_1} \right) \sin^2 \theta \cos^2 \theta + \frac{1}{E_2} \cos^4 \theta \right]^{-1} \tag{8.17}
$$

$$
G_{xy} = \left[\frac{1}{G_{12}} + 4 \left(\frac{1}{E_1} + \frac{1}{E_2} + \frac{2\nu_{12}}{E_1} - \frac{1}{G_{12}} \right) \sin^2 \theta \cos^2 \theta \right]^{-1}
$$

$$
\eta_{x,xy} = E_x \left[\left(\frac{2}{E_1} + \frac{2\nu_{12}}{E_1} - \frac{1}{G_{12}} \right) \sin \theta \cos^3 \theta \right.
$$
$$
\left. - \left(\frac{2}{E_2} + \frac{2\nu_{12}}{E_1} - \frac{1}{G_{12}} \right) \sin^3 \theta \cos \theta \right]
$$

$$\eta_{y,xy} = E_y \left[\left(\frac{2}{E_1} + \frac{2\nu_{12}}{E_1} - \frac{1}{G_{12}} \right) \sin^3 \theta \cos \theta \right.$$
$$\left. - \left(\frac{2}{E_2} + \frac{2\nu_{12}}{E_1} - \frac{1}{G_{12}} \right) \sin \theta \cos^3 \theta \right]$$

Variations of the apparent moduli E_x, G_{xy}, ν_{xy}, and $\eta_{x,xy}$ against fiber orientation θ for three composites are given in Fig. 8.2.

In the plots, the following material constants were used.

Carbon/epoxy:

$$E_1 = 140 \text{ GPa}, \qquad E_2 = 10 \text{ GPa}$$
$$G_{12} = 7.0 \text{ GPa}, \qquad \nu_{12} = 0.3$$

Boron/aluminum:

$$E_1 = 235 \text{ GPa}, \qquad E_2 = 135 \text{ GPa}$$
$$G_{12} = 45 \text{ GPa}, \qquad \nu_{12} = 0.3$$

Glass/epoxy:

$$E_1 = 43 \text{ GPa}, \qquad E_2 = 9.0 \text{ GPa}$$
$$G_{12} = 4.5 \text{ GPa}, \qquad \nu_{12} = 0.27$$

Among the three composites, carbon/epoxy is the most anisotropic with the largest E_1/E_2 ratio, and boron/aluminum is the least anisotropic. The following behaviors are noticed.

- The longitudinal stiffness represented by E_x drops sharply as the loading direction deviates from the fiber direction, expecially for carbon/epoxy. Thus, the fiber orientation in composite structures must be precisely aligned during manufacturing.
- The shear stiffness represented by G_{xy} attains a maximum value at $\theta = 45°$. This means that placing fibers in the 45° direction in a composite can achieve the best in-plane shear property.
- The maximum coupling between extension and shear occurs between $\theta = 10°$ and $20°$. From the results for $\eta_{x,xy}$ shown in Fig. 8.2, a unit of axial strain ($\varepsilon_x = 1$) applied in the x-direction can induce two units of shear strain γ_{xy} in carbon/epoxy composite if the fiber orientation is $12°$.

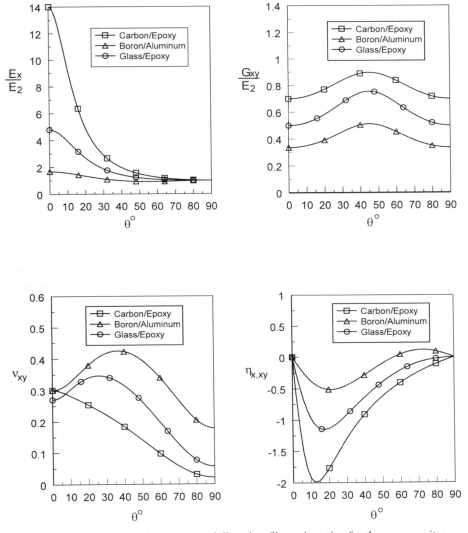

Figure 8.2   Variations of apparent moduli against fiber orientation for three composites.

8.2   OFF-AXIS LOADING

Consider a state of uniform deformation in a composite panel produced by applying a uniaxial stress $\sigma_{xx} = \sigma_0$ in the x-direction (see Fig. 8.3). From (8.12), the uniform state of deformation is given by the strains

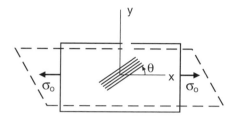

Figure 8.3   Deformation of an off-axis composite under tension.

$$\varepsilon_{xx} = \bar{S}_{11}\sigma_0$$
$$\varepsilon_{yy} = \bar{S}_{12}\sigma_0 \qquad (8.18)$$
$$\gamma_{xy} = \bar{S}_{16}\sigma_0$$

It is seen that shear deformation can result from application of a normal load, except when x and y axes coincide with the material principal axes, x_1 and x_2 (in that case $\bar{S}_{16} = S_{16} = 0$). Coupling between normal deformation and shear deformation does not exist in isotropic solids.

Integrating the strain–displacement relations for ε_{xx} and ε_{yy} [see (2.7) and (2.8)] yields the displacement components:

$$u = \varepsilon_{xx}x + f(y) \qquad (8.19)$$
$$v = \varepsilon_{yy}y + g(x) \qquad (8.20)$$

where $f(y)$ and $g(x)$ are arbitrary functions of variables y and x, respectively. Substituting (8.19) and (8.20) into shear strain

$$\gamma_{xy} = \frac{\partial u}{\partial y} + \frac{\partial v}{\partial x} \qquad (8.21)$$

we obtain

$$\gamma_{xy} = f'(y) + g'(x) = \bar{S}_{16}\sigma_0 \qquad (8.22)$$

where a prime indicates differentiation with respect to the argument. From (8.22), it is obvious that $f(y)$ and $g(x)$ must be linear functions of y and x, respectively, i.e.,

$$f(y) = C_1 y + C_3 \qquad (8.23)$$
$$g(x) = C_2 x + C_4 \qquad (8.24)$$

Thus, the displacements (8.19) and (8.20) can be expressed as

$$u = \bar{S}_{11}\sigma_0 x + C_1 y + C_3 \tag{8.25}$$

$$v = \bar{S}_{12}\sigma_0 y + C_2 x + C_4 \tag{8.26}$$

Removing the rigid body translations from the displacements above, we set $C_3 = C_4 = 0$. To suppress the rigid body rotation, we assume that the horizontal edges of the panel remain horizontal after deformation, i.e.,

$$\frac{\partial v}{\partial x} = C_2 = 0 \tag{8.27}$$

The remaining constant C_1 is obtained from (8.22) in conjunction with (8.21). We have

$$C_1 = \bar{S}_{16}\sigma_0 \tag{8.28}$$

Thus, the displacement field in the composite panel under the uniform stress $\sigma_{xx} = \sigma_0$ is

$$u = \bar{S}_{11}\sigma_0 x + \bar{S}_{16}\sigma_0 y \tag{8.29}$$

$$v = \bar{S}_{12}\sigma_0 y \tag{8.30}$$

For the AS4/3501-6 carbon/epoxy composite, the elastic moduli are

$$\begin{aligned} E_1 &= 140 \text{ GPa} \\ E_2 &= 10 \text{ GPa} \\ G_{12} &= 6.9 \text{ GPa} \\ \nu_{12} &= 0.3 \end{aligned} \tag{8.31}$$

If the off-axis angle θ is $45°$, then

$$\begin{aligned} \bar{S}_{11} &= 0.615 \times 10^{-10} \text{ m}^2/\text{N} \\ \bar{S}_{12} &= -0.245 \times 10^{-10} \text{ m}^2/\text{N} \\ \bar{S}_{16} &= -0.47 \times 10^{-10} \text{ m}^2/\text{N} \end{aligned} \tag{8.32}$$

The deformed shape of the panel can be determined from the displacement field given by (8.29) and (8.30), which is depicted in Fig. 8.3.

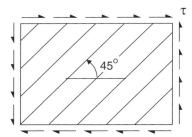

Figure 8.4   Composite panel under shear load.

Example 8.1   Off-Axis Composite Panel under Shear Load   A composite panel with $\theta = 45°$ is under shear load τ, as shown in Fig. 8.4. The elastic moduli of the composite are given by (8.31). From (8.12), the strains produced by the shear load are

$$\varepsilon_{xx} = \bar{S}_{16}\tau \tag{a}$$

$$\varepsilon_{yy} = \bar{S}_{26}\tau \tag{b}$$

$$\gamma_{xy} = \bar{S}_{66}\tau \tag{c}$$

Using the elastic moduli (8.31) and the transformed compliances (8.14), we have

$$\bar{S}_{16} = \bar{S}_{26} = -0.47 \times 10^{-10} \ \mathrm{m^2/N} \tag{d}$$

$$\bar{S}_{66} = 1.45 \times 10^{-10} \ \mathrm{m^2/N} \tag{e}$$

It is interesting to note from (d) that a positive shear load produces negative normal strains in ε_{xx} and ε_{yy}. In other words, a positive shear load would shorten the composite panel in both x and y directions. On the other hand, a negative shear load would enlarge the size of the panel.

8.3   NOTATION FOR STACKING SEQUENCE IN LAMINATES

Unidirectionally reinforced fiber composites have superior properties only in the fiber direction. In practical applications, laminas with various fiber orientations are combined together to form laminated composites that are capable of carrying loads of multiple directions. Due to the lamination, the material properties of a laminate become heterogeneous over the thickness. Further, due to the arbitrary fiber orientations of the laminas, the laminate may not possess orthotropy as each constituent lamina does.

A laminate consists of a number of laminas of different fiber orientations. A composite ply is the basic element in constructing a laminate. Each lamina may contain one or more plies of the same fiber orientation. The laminate properties depend on the lamina fiber orientation as well as its position in the laminate (the **stacking sequence**). To describe a laminate, the fiber orientation and position of each lamina must be accurately specified.

To achieve the foregoing purpose, a global coordinate system, (x, y, z), must be established. Let the $x-y$ plane be parallel to the plane of the laminate and the z-axis be in the thickness direction. The fiber orientation (θ) is measured relative to the x-axis as shown in Fig. 8.1. The positions of the plies are listed in sequence starting from one face of the laminate to the other face along the positive z-direction. An example is shown in Fig. 8.5 for $[0/0/45/-45]$.

In practice, layup is not arbitrary; it often possesses certain repetitions and symmetry. To avoid lengthy expressions, abbreviated notations are used to specify the stacking sequence. The following are some abbreviated notations introduced to indicate ply or sublaminate repetition and symmetry in the layup.

Symmetry If the layup is symmetric with respect to the midplane of the laminate, then only half of the plies are specified; the other half is included by a subscript s indicating symmetric layup. An example is $[0/90/+45/-45]_s$ or $[0/90/\pm45]_s$ which stands for $[0/90/+45/-45/-45/+45/90/0]$.

Repetition If a ply or sublaminate contiguously repeats itself n times in a laminate, then a subscript n is attached to the ply angle or the sublaminate group angles to indicate the repetitions. For example, $[0_2/90_2]$ stands for $[0/0/90/90]$, and $[(0/90)_2/\pm45_2]$ stands for $[0/90/0/90/45/45/-45/-45]$.

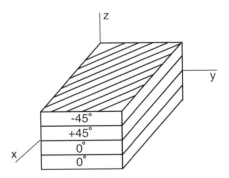

Figure 8.5 Position of plies in a laminate.

Additional examples are:

$[0/\pm45/\underline{0}]_s$ stands for $[0/45/-45/0/-45/45/0]$ where an underline is used to indicate the ply right on the plane of symmetry.

$[(0/90)_2]_s$ or $[0/90]_{2s}$ stands for $[0/90/0/90/90/0/90/0]$.

$[(0/90)_s]_2$ stands for $[0/90/90/0/0/90/90/0]$.

Special names are often given to laminates with particular layups. For instance, a laminate consisting of only $0°$ and $90°$ plies is referred to as a **cross-ply laminate**. If a laminate consists of only $+\theta$ and $-\theta$ plies, it is called an **angle-ply laminate**. A **balanced laminate** is a laminate in which there is a $-\theta$ ply for every $+\theta$ ply.

8.4 SYMMETRIC LAMINATE UNDER IN-PLANE LOADING

Laminates with symmetric lay-ups are most popular in applications mainly because they are free from warping induced by thermal residual stresses resulting from curing at elevated temperatures. Laminates provide excellent stiffness and strength properties for in-plane loading. They can be used with great structural efficiency in skins and stiffeners in aircraft structures.

Consider a laminated panel consisting of a number of fiber-reinforced laminas, as shown in Fig. 8.6. The $x-y$ plane is located at the midplane of the laminate. In a well-made laminate, the laminas are perfectly bonded. Under in-plane loading, laminas deform, and the displacements are continuous across the ply boundaries. If the laminate is thin, then we can assume with good accuracy that the strain components ε_{xx}, ε_{yy}, and γ_{xy} in all the laminas are the same over the thickness of the laminate. In other words, we assume that the laminate deforms uniformly over the thickness if it is under in-plane loading.

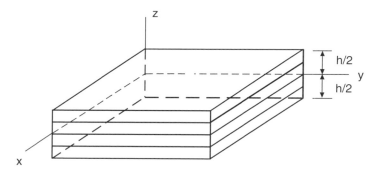

Figure 8.6 Laminated panel.

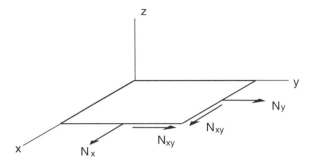

Figure 8.7 Resultant forces.

Although the strains are uniform and continuous over the thickness of the laminate, the stresses in the laminas are, in general, discontinuous across the interfaces due to different material properties resulting from different fiber orientations. For the kth lamina, the stress components are calculated using the transformed stress-strain relations

$$\left\{ \begin{array}{c} \sigma_{xx} \\ \sigma_{yy} \\ \sigma_{xy} \end{array} \right\}_k = \left[\begin{array}{ccc} \bar{Q}_{11} & \bar{Q}_{12} & \bar{Q}_{16} \\ \bar{Q}_{12} & \bar{Q}_{22} & \bar{Q}_{26} \\ \bar{Q}_{16} & \bar{Q}_{26} & \bar{Q}_{66} \end{array} \right]_k \left\{ \begin{array}{c} \varepsilon_{xx} \\ \varepsilon_{yy} \\ \gamma_{xy} \end{array} \right\} \tag{8.33}$$

It is conceivable that analyzing each layer individually is a cumbersome task. Consistent with the beam theory in Chapter 4 and the plate theory in Chapter 7, in-plane resultant forces are introduced. These resultant forces are defined as

$$\left\{ \begin{array}{c} N_x \\ N_y \\ N_{xy} \end{array} \right\} = \int_{-h/2}^{h/2} \left\{ \begin{array}{c} \sigma_{xx} \\ \sigma_{yy} \\ \sigma_{xy} \end{array} \right\} dz \tag{8.34}$$

where h denotes the thickness of the plate and the stress components σ_{xx}, σ_{yy}, and σ_{xy} assume the values of $\sigma_{xx}^{(k)}$, $\sigma_{yy}^{(k)}$, and $\sigma_{xy}^{(k)}$ if z is located in the kth lamina. The resultant forces, which have the unit of force per unit length, are depicted in Fig. 8.7.

Let the kth lamina occupy the region from $z = z_{k-1}$ to $z = z_k$ (see Fig. 8.8). Then the integral in (8.34) can be broken up into integrals over the individual laminas as

$$\left\{ \begin{array}{c} N_x \\ N_y \\ N_{xy} \end{array} \right\} = \sum_{k=1}^{n} \int_{z_{k-1}}^{z_k} \left\{ \begin{array}{c} \sigma_{xx} \\ \sigma_{yy} \\ \sigma_{xy} \end{array} \right\}_k dz \tag{8.35}$$

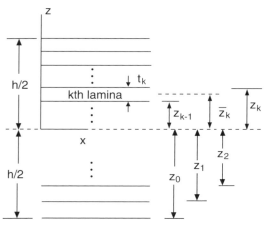

Figure 8.8   Positions of laminas.

where n is the total number of laminas in the laminate. Substituting (8.33) into (8.35), we obtain

$$\left\{ \begin{array}{c} N_x \\ N_y \\ N_{xy} \end{array} \right\} = \sum_{k=1}^{n} \int_{z_{k-1}}^{z_k} \left[\bar{Q} \right]_k \left\{ \begin{array}{c} \varepsilon_{xx} \\ \varepsilon_{yy} \\ \gamma_{xy} \end{array} \right\} dz \tag{8.36}$$

Note that the quantities ε_x, ε_y, and γ_{xy} are independent of z. Hence, the integrations in (8.36) can be performed to yield

$$\left\{ \begin{array}{c} N_x \\ N_y \\ N_{xy} \end{array} \right\} = \left[\begin{array}{ccc} A_{11} & A_{12} & A_{16} \\ A_{12} & A_{22} & A_{26} \\ A_{16} & A_{26} & A_{66} \end{array} \right] \left\{ \begin{array}{c} \varepsilon_{xx} \\ \varepsilon_{yy} \\ \gamma_{xy} \end{array} \right\} \tag{8.37}$$

where

$$A_{ij} = \int_{-h/2}^{h/2} \bar{Q}_{ij}^{(k)} \, dz \qquad (i, j = 1, 2, 6) \tag{8.38}$$

or, explicitly,

$$A_{ij} = \sum_{k=1}^{n} \bar{Q}_{ij}^{(k)} (z_k - z_{k-1}) = \sum_{k=1}^{n} \bar{Q}_{ij}^{(k)} t_k \tag{8.39}$$

Coefficients A_{ij} are called **extensional stiffnesses**.

8.5   EFFECTIVE MODULI FOR SYMMETRIC LAMINATES

A symmetric laminate under in-plane loading can be treated as an equivalent homogeneous anisotropic solid in plane stress by introducing the average stresses

$$\bar{\sigma}_{xx} = \frac{N_x}{h}, \qquad \bar{\sigma}_{yy} = \frac{N_y}{h}, \qquad \bar{\sigma}_{xy} = \frac{N_{xy}}{h} \qquad (8.40)$$

In terms of these average stresses, (8.37) can be written as

$$\left\{ \begin{array}{c} \bar{\sigma}_{xx} \\ \bar{\sigma}_{yy} \\ \bar{\sigma}_{xy} \end{array} \right\} = \frac{1}{h} [A] \left\{ \begin{array}{c} \varepsilon_{xx} \\ \varepsilon_{yy} \\ \gamma_{xy} \end{array} \right\} \qquad (8.41)$$

Equation (8.41) indicates that the laminate is effectively a 2-D anisotropic solid inplane stress and $[A]/h$ is the effective elastic constant matrix.

The inverse relation of (8.41) is

$$\{\varepsilon\} = h \left[A' \right] \{\bar{\sigma}\} \qquad (8.42)$$

where

$$[A'] = [A]^{-1}$$

The components A'_{ij} are given by

$$A'_{11} = \frac{A_{22}A_{66} - A_{26}^2}{\Delta}$$

$$A'_{12} = \frac{A_{16}A_{26} - A_{12}A_{66}}{\Delta}$$

$$A'_{22} = \frac{A_{11}A_{66} - A_{16}^2}{\Delta}$$

$$A'_{16} = \frac{A_{12}A_{26} - A_{22}A_{16}}{\Delta} \qquad (8.43)$$

$$A'_{26} = \frac{A_{12}A_{16} - A_{11}A_{26}}{\Delta}$$

$$A'_{66} = \frac{A_{11}A_{22} - A_{12}^2}{\Delta}$$

where

$$\Delta = \left| A_{ij} \right|$$

Comparing (8.42) with (8.15), we can relate the components A'_{ij} to the effective engineering moduli for the laminate as

$$E_x = \frac{1}{hA'_{11}}, \qquad E_y = \frac{1}{hA'_{22}}$$

$$\nu_{xy} = -\frac{A'_{12}}{A'_{11}}, \qquad \nu_{yx} = -\frac{A'_{12}}{A'_{22}} \qquad (8.44)$$

$$\eta_{xy,x} = \frac{A'_{16}}{A'_{66}}, \qquad \eta_{xy,y} = \frac{A'_{26}}{A'_{66}}$$

$$G_{xy} = \frac{1}{hA'_{66}}$$

If a symmetric laminate also possesses the property $A_{16} = A_{26} = 0$ (e.g., $[0/90]_s$ and $[\pm 45]_s$), then the effective moduli can be expressed explicitly as

$$E_x = \frac{A_{11}A_{22} - A_{12}^2}{hA_{22}}$$

$$E_y = \frac{A_{11}A_{22} - A_{12}^2}{hA_{11}}$$

$$\nu_{xy} = \frac{A_{12}}{A_{22}} \qquad (8.45)$$

$$\nu_{yx} = \frac{A_{12}}{A_{11}}$$

$$G_{xy} = \frac{A_{66}}{h}$$

$$\eta_{xy,x} = \eta_{xy,y} = 0$$

Quasi-isotropic Laminate   Although each lamina is anisotropic, some laminates may possess isotropic in-plane stiffness properties. These laminates are called **quasi-isotropic laminates** because their bending properties are still anisotropic.

A quasi-isotropic laminate is characterized by an in-plane stiffness matrix $[A]$ that is invariant with respect to in-plane rotations of the coordinate system. In other words, the laminate stiffnesses are independent of direction just like an isotropic homogeneous solid. A symmetric quasi-isotropic laminate is constructed as follows. Let n (≥ 3) be the number of fiber orientations in the laminate. The angle between two adjacent fiber orientations is π/n. The number of plies for each fiber orientation is identical. Examples for quasi-isotropic laminates are $\pi/3$ laminate $[0/\pm 60]_s$ and $\pi/4$ laminate

$[\pm 45/0/90]_s$. Other quasi-isotropic laminates corresponding to higher values of n can be constructed in the same manner.

Example 8.2   Effective Moduli for Quasi-isotropic Laminate $[\pm 45/0/90]_s$
Assume that the elastic moduli of the composite are

$$E_1 = 180 \text{ GPa}, \qquad E_2 = 10 \text{ GPa}, \qquad G_{12} = 7 \text{ GPa}, \qquad \nu_{12} = 0.3$$

$$\text{ply thickness} = 0.127 \text{ mm}$$
$$\text{laminate thickness } h = 1.02 \text{ mm}$$

The laminate in-plane stiffnesses A_{ij} are obtained as

$$A_{11} = A_{22} = 61.84 \times 10^6 \text{ N/m}$$

$$A_{12} = A_{21} = 17.92 \times 10^6 \text{ N/m}$$

$$A_{16} = A_{26} = 0$$

$$A_{66} = 21.96 \times 10^6 \text{ N/m}$$

The effective moduli for the laminate are calculated using (8.45). We obtain

$$E_x = E_y = \frac{A_{11}A_{22} - A_{12}^2}{hA_{22}} = 55.5 \text{ GPa}$$

$$\nu_{xy} = \nu_{yx} = \frac{A_{12}}{A_{22}} = 0.29$$

$$G_{xy} = \frac{A_{66}}{h} = 21.5 \text{ GPa}$$

Note that the relation $E = 2(1 + \nu)G$ for isotropic materials holds for the quasi-isotropic laminate.

Example 8.3   Negative Poisson's Ratio in Laminates   When a symmetric laminate is treated as a 2-D homogeneous solid in plane stress, it may exhibit some unusual properties that are not observed in other homogeneous solids. One of these is a negative Poisson's ratio.

Consider the symmetric but unbalanced laminates $[\theta/\theta + 25°]_s$, where $\theta = 0°$ to $180°$. The ply properties are

$$E_1 = 180 \text{ GPa}, \qquad E_2 = 10 \text{ GPa}, \qquad G_{12} = 7 \text{ GPa}$$

$$\nu_{12} = 0.28, \quad \text{ply thickness} = 0.13 \text{ mm}$$

The apparent Poisson's ratio ν_{xy} can be calculated using (8.44). Figure 8.9 shows ν_{xy} as a function of θ. It is seen that negative values of ν_{xy} are possible. Also note that unusually high positive values of ν_{xy} can be produced.

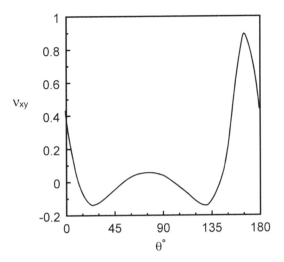

Figure 8.9   Poisson's ratio v_{xy} as a function of fiber orientation θ in $[\theta/\theta + 25°]_s$ laminate.

8.6   LAMINAR STRESSES

The in-plane resultant forces provide a convenient way to formulate the global governing equations for thin laminates. However, for the prediction of laminate strength, stresses in each lamina must be recovered.

If loads in terms of $\{N\}$ are given, then the laminate strains are obtained from

$$\{\varepsilon\} = [A]^{-1}\{N\} \qquad (8.46)$$

The stresses $\sigma_{xx}^{(k)}$, $\sigma_{yy}^{(k)}$, and $\sigma_{xy}^{(k)}$ in the kth lamina are calculated according to (8.33), i.e.,

$$\{\sigma\}_k = [\bar{Q}]_k \{\varepsilon\} \qquad (8.47)$$

Example 8.4   Laminar Stresses in a Quasi-isotropic Symmetric Laminate
$[\pm45/0/90]_s$ under a Uniaxial Load N_x   The composite ply properties are assumed to be

$$E_1 = 140 \text{ GPa}, \qquad E_2 = 10 \text{ GPa} \qquad (a)$$

$$G_{12} = 7 \text{ GPa}, \qquad v_{12} = 0.3$$

ply thickness = 0.127 mm

The following stiffness matrices are readily calculated:

$$[\bar{Q}]_{0°} = \begin{bmatrix} 140.9 & 3.02 & 0 \\ 3.02 & 10.06 & 0 \\ 0 & 0 & 7.0 \end{bmatrix} \times 10^9 \text{ Pa} \qquad\qquad\text{(b)}$$

$$[\bar{Q}]_{90°} = \begin{bmatrix} 10.06 & 3.02 & 0 \\ 3.02 & 140.9 & 0 \\ 0 & 0 & 7.0 \end{bmatrix} \times 10^9 \text{ Pa} \qquad\qquad\text{(c)}$$

$$[\bar{Q}]_{\pm45°} = \begin{bmatrix} 46.25 & 32.25 & \pm32.71 \\ 32.25 & 46.25 & \pm32.71 \\ \pm32.71 & \pm32.71 & 36.23 \end{bmatrix} \times 10^9 \text{ Pa} \qquad\text{(d)}$$

$$[A] = \begin{bmatrix} 61.84 & 17.92 & 0 \\ 17.92 & 61.84 & 0 \\ 0 & 0 & 21.96 \end{bmatrix} \times 10^6 \text{ N/m} \qquad\qquad\text{(e)}$$

$$[A]^{-1} = \begin{bmatrix} 17.65 & -5.11 & 0 \\ -5.11 & 17.65 & 0 \\ 0 & 0 & 45.53 \end{bmatrix} \times 10^{-9} \text{ m/N} \qquad\text{(f)}$$

The strains are

$$\begin{Bmatrix} \varepsilon_{xx} \\ \varepsilon_{yy} \\ \gamma_{xy} \end{Bmatrix} = [A]^{-1} \begin{Bmatrix} N_x \\ 0 \\ 0 \end{Bmatrix} = \begin{Bmatrix} 17.65 \\ -5.11 \\ 0 \end{Bmatrix} \times 10^{-9} \, N_x$$

in which N_x is in N/m. The laminar stresses are

$$\begin{Bmatrix} \sigma_{xx} \\ \sigma_{yy} \\ \sigma_{xy} \end{Bmatrix}_{0°} = [\bar{Q}]_{0°}\{\varepsilon\} = \begin{Bmatrix} 2472 \\ 1.8 \\ 0 \end{Bmatrix} N_x \qquad \text{N/m}^2$$

$$\begin{Bmatrix} \sigma_{xx} \\ \sigma_{yy} \\ \sigma_{xy} \end{Bmatrix}_{90°} = [\bar{Q}]_{90°}\{\varepsilon\} = \begin{Bmatrix} 162.2 \\ -667.3 \\ 0 \end{Bmatrix} N_x \qquad \text{N/m}^2$$

$$\begin{Bmatrix} \sigma_{xx} \\ \sigma_{yy} \\ \sigma_{xy} \end{Bmatrix}_{\pm45°} = [\bar{Q}]_{\pm45°}\{\varepsilon\} = \begin{Bmatrix} 651.5 \\ 332.8 \\ \pm410.1 \end{Bmatrix} N_x \qquad \text{N/m}^2$$

The distributions of the laminar normal stresses are shown in Fig. 8.10. It is evident that the stress distribution from lamina to lamina is not continuous. Also note the large compressive stress σ_{yy} developed in the 90° lamina due to its inability to contract in the y-direction. It is easy to verify that the resultant force $N_y = 0$.

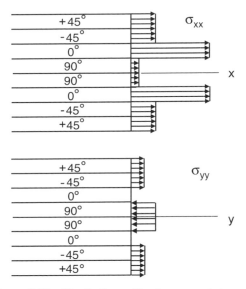

Figure 8.10   Distributions of laminar normal stresses.

The stress components in reference to the material principal axes (x_1, x_2) in each lamina are of particular interest in strength analysis. These stress components can be obtained using the coordinate transformation given by (8.5). We obtain

$$\left\{ \begin{array}{c} \sigma_{11} \\ \sigma_{22} \\ \sigma_{12} \end{array} \right\}_{0°} = \left\{ \begin{array}{c} 2472 \\ 1.8 \\ 0 \end{array} \right\} N_x \qquad \text{N/m}^2$$

$$\left\{ \begin{array}{c} \sigma_{11} \\ \sigma_{22} \\ \sigma_{12} \end{array} \right\}_{90°} = \left\{ \begin{array}{c} -667.3 \\ 162.2 \\ 0 \end{array} \right\} N_x \qquad \text{N/m}^2$$

$$\left\{ \begin{array}{c} \sigma_{11} \\ \sigma_{22} \\ \sigma_{12} \end{array} \right\}_{\pm45°} = \left\{ \begin{array}{c} 902.2 \\ 82.0 \\ \mp159.3 \end{array} \right\} N_x \qquad \text{N/m}^2$$

8.7   [±45°] LAMINATE

The ±45° type of laminate is often used to provide greater shear rigidities of composite structures. For example, consider the $[\pm45]_s$ laminate with the composite properties given in Example 8.4. Using $\left[\bar{Q} \right]_{\pm45°}$ given by (d) in

Example 8.4, we have

$$[A] = \begin{bmatrix} 23.50 & 16.38 & 0 \\ 16.38 & 23.50 & 0 \\ 0 & 0 & 18.40 \end{bmatrix} \times 10^6 \text{ N/m} \qquad (8.48)$$

By treating this laminate as an equivalent homogeneous orthotropic solid in plane stress, the equivalent elastic moduli can be obtained from (8.45). We have

$$E_x = E_y = 23.8 \text{ GPa}, \qquad G_{xy} = 36.2 \text{ GPa}$$
$$\nu_{xy} = \nu_{yx} = 0.70, \qquad \eta_{x,xy} = \eta_{xy,x} = 0 \qquad (8.49)$$

Comparing these moduli with those of the unidirectional composite, we note a significant increase in the shear rigidity. However, this is achieved at the expense of the longitudinal modulus E_x.

Determination of G_{12} Using ±45° Laminates   Consider the $[\pm 45]_s$ laminate subjected to a uniform uniaxial stress $\bar{\sigma}_{xx}(= N_x/h) = \sigma_0$ and $\bar{\sigma}_{yy} = \bar{\sigma}_{xy} = 0$. Based on the symmetry of the laminate, it is not difficult to see that in both $+45°$ and $-45°$ laminas,

$$\sigma_{xx} = \sigma_0$$
$$\sigma_{yy} = 0 \qquad (8.50)$$
$$\left(\sigma_{xy}\right)_{-45°} = -\left(\sigma_{xy}\right)_{+45°}$$

From the coordinate transformation (8.5) for stress, we have

$$\sigma_{12} = -\sin\theta\cos\theta\,\sigma_{xx} + \sin\theta\cos\theta\,\sigma_{yy} + (\cos^2\theta - \sin^2\theta)\,\sigma_{xy}$$
$$= -\sin\theta\cos\theta\,\sigma_0 + \left(\cos^2\theta - \sin^2\theta\right)\sigma_{xy} \qquad (8.51)$$

For $\theta = -45°$, (8.51) yields

$$\sigma_{12} = \tfrac{1}{2}\sigma_0 \qquad (8.52)$$

Since the $[\pm 45]_s$ laminate is balanced with respect to the uniaxial load, we have $\gamma_{xy} = 0$. From the coordinate transformation on strains,

$$\left\{ \begin{array}{c} \varepsilon_{11} \\ \varepsilon_{22} \\ \gamma_{12} \end{array} \right\} = [T_\varepsilon] \left\{ \begin{array}{c} \varepsilon_{xx} \\ \varepsilon_{yy} \\ \gamma_{xy} \end{array} \right\} \tag{8.53}$$

we obtain

$$\gamma_{12} = -2 \sin\theta \cos\theta (\varepsilon_{xx} - \varepsilon_{yy}) \tag{8.54}$$

In the $-45°$ lamina, (8.54) gives

$$\gamma_{12} = \varepsilon_{xx} - \varepsilon_{yy} \tag{8.55}$$

With respect to the material principal coordinate system (x_1, x_2), the shear stress-strain relation for an orthotropic composite is $\sigma_{12} = G_{12}\gamma_{12}$. Thus,

$$G_{12} = \frac{\sigma_{12}}{\gamma_{12}} = \frac{\sigma_{12}}{\varepsilon_{xx} - \varepsilon_{yy}} \tag{8.56}$$

In view of (8.52), we conclude that

$$G_{12} = \frac{\sigma_0}{2(\varepsilon_{xx} - \varepsilon_{yy})} = \frac{E_x}{2(1 + \nu_{xy})} \tag{8.57}$$

In deriving the relation above, the definitions $E_x = \sigma_0/\varepsilon_{xx}$ and $\nu_{xy} = -\varepsilon_{yy}/\varepsilon_{xx}$ have been used. Here E_x and ν_{xy} are the effective Young's modulus and Poisson's ratio, respectively, of the $[\pm45]_s$ laminate under uniaxial load in the x-direction. The relation (8.57) can be used to determine G_{12} from the tension test of a $[\pm45]_s$ laminate specimen.

PROBLEMS

8.1 Given a carbon/epoxy composite panel under uniaxial loading, i.e., $\sigma_{xx} = \sigma_0$, $\sigma_{yy} = \sigma_{xy} = 0$, plot γ_{xy} as a function of the fiber orientation θ. The composite properties are

$$E_1 = 140\,\text{GPa}, \qquad E_2 = 10\,\text{GPa}, \qquad G_{12} = 7\,\text{GPa}, \qquad \nu_{12} = 0.3$$

8.2 Consider a rectangular composite panel with $\theta = 45°$ (material properties are given in Problem 8.1) subjected to $\sigma_{xx} = 10$ MPa, $\sigma_{yy} = 0$, $\sigma_{xy} = \tau$. Find τ that is necessary to keep the deformed shape rectangular.

8.3 Plot the extension-shear coupling coefficients $\eta_{x,xy}$ and $\eta_{xy,x}$ versus θ for the composite given in Problem 8.1. Find the θ's that correspond to the maximum values of $\eta_{x,xy}$ and $\eta_{xy,x}$, respectively.

8.4 If the carbon/epoxy composite panel is subjected to a shear stress τ_{xy}, find

(a) the fiber orientation at which σ_{11} is maximum

(b) the fiber orientation at which γ_{xy} is minimum.

Compare the result with that of Problem 8.3.

8.5 Consider a $[\pm 45]_s$ laminate. If the constituent composite material is highly anisotropic, i.e.,

$$E_1 \gg E_2 \quad \text{and} \quad E_1 \gg G_{12}$$

show that the effective engineering moduli for the laminate can be expressed approximately as

$$E_x \simeq 4Q_{66} \simeq 4G_{12}$$

$$G_{xy} \simeq \frac{Q_{11}}{4} \simeq \frac{E_1}{4}$$

$$v_{xy} \simeq \frac{Q_{11} - 4Q_{66}}{Q_{11} + 4Q_{66}} \simeq \frac{E_1 - 4G_{12}}{E_1 + 4G_{12}}$$

Compare these approximate values with the exact values for AS4/3501-6 carbon/epoxy composite.

8.6 Compare the in-plane longitudinal stiffnesses in the x-direction for $[\pm 30/0]_s$ and $[30_2/0]_s$ laminates of AS4/3501-6 carbon/epoxy composite. Which is stiffer?

8.7 Plot the effective moduli E_x, G_{xy}, and v_{xy} versus θ for the angle-ply laminate $[\pm\theta]_s$ of AS4/3501-6 carbon/epoxy composite.

8.8 Find the shear strains (γ_{xy}) in the AS4/3501-6 carbon/epoxy composite $[\pm 45]_s$ and $[0/90]_s$ laminates subjected to the shear loading $N_{xy} = 1000$ N/m. Also find the lamina stresses σ_{11}, σ_{22}, and σ_{12}. If the maximum shear strength of the composite is $|\sigma_{12}| = 100$ MPa, what are the shear loads (N_{xy}) the two laminates can carry?

INDEX